Silicon-on-Insulator:

Its Technology and Applications

Silicon-on-Insulator: Its Technology and Applications

Edited by S. Furukawa

KTK Scientific Publishers / Tokyo

D. Reidel Publishing Company / Dordrecht, Boston, London

A MEMBER OF THE
KLUWER ACADEMIC PUBLISHERS GROUP

Library of Congress Cataloging in Publication Data

Main entry under title:

Silicon-on-insulator.

"Papers presented in US-Japan Seminar on 'Solid Phase
Epitaxy and Interface Kinetics' held in Ōiso, Japan,
June 20-24, 1983 ... co-sponsored by National Science
Foundation and Japan Society for the Promotion of
Science"--P.
    Companion volume containing the other papers presented
at seminar published under title: Layered structures
and interface kinetics.
    "ASST, advances in solid state technology."
    1. Solid state electronics--Congresses. 2. Integrated
circuits--Design and construction--Congresses.
3. Silicon--Congresses. 4. Epitaxy--Congresses.
I. Furukawa, S. (Seijirō), 1934-   . II. US-Japan
Seminar on "Solid Phase Epitaxy and Interface Kinetics"
(1983 : Ōiso-machi, Japan) III. National Science
Foundation (U.S.) IV. Nihon Gakujutsu Shinkōkai.
V. Layered structures and interface kinetics.
TK7871.85.S548   1984   621.381   84-27751

ISBN-13: 978-94-010-8846-6   e-ISBN-13: 978-94-009-5311-6
DOI: 10.1007/978-94-009-5311-6

Published by KTK Scientific Publishers (KTK),
307 Shibuyadai-haim, 4-17 Sakuragaoka-cho, Shibuya-ku, Tokyo 150, Japan,
in co-publication with D. Reidel Publishing Company, Dordrecht, Holland

Sold and distributed in the U.S.A. and Canada
by Kluwer Academic Publishers,
190 Old Derby Street, Hingham, MA 02043, U.S.A.,
in Japan by KTK Scientific Publishers (KTK),
307 Shibuyadai-haim, 4-17 Sakuragaoka-cho, Shibuya-ku, Tokyo 150, Japan

In all other countries, sold and distributed
by Kluwer Academic Publishers Group,
P.O. Box 322, 3300 AH Dordrecht, Holland

Preface

This volume contains papers presented during the US-Japan seminar on "Solid Phase Epitaxy and Interface Kinetics" held in Oiso, Japan, June 20-24, 1983. This seminar was co-sponsored by the National Science Foundation and Japan Society for the Promotion of Science and co-chaired by Professor S. Furukawa, Tokyo Insitute of Technology, and Professor J. W. Mayer, Cornell University. Extensive topics such as solid phase epitaxy, growth mechanisms and interface kinetics, silicon-on-insulator structures, silicide-on-Si sturctures, novel nanometer and layered devices, and so on were discussed and more than 50 papers were presented. Most papers were original ones with brief reviews added for the convenience of the readers at the editor's request.

The editor classified these papers into two groups and compiled two volumes; "Silicon-on-Insulator (SOI): Its Technology and Applications" and "Layered Structures and Interface Kinetics: Their Technology and Applications". This volume mainly contains the papers related to epitaxial growth of metal, insulator and semiconductor films, growth mechanisms, interface kinetics, properties and applications of silicide films, and novel nanometer and layered devices. These papers offer basic properties of the layered structures and possibility of various applications of the strucures to present and future semiconductor devices.

The editor is indebted to our fellow contributors who agreed to partake in publishing the proceedings of the seminar, to Japanese principal participants of the seminar for encouraging him to have the seminar and to compile these volumes, to Professor H. Ishiwara for his secretarial work throughout the seminar and the publication.

October 1, 1984

Seijiro Furukawa

CONTENTS

CHAPTER 3: SOLID PHASE EPITAXY

CHAPTER 4: CHARACTERIZATION AND DEVICE APPLICATIONS

CHAPTER 1 : LASER AND ELECTRON-BEAM RECRYSTAL-LIZATION

Silicon-on-Insulator: Its Technology and Applications, edited by S. Furukawa, pp. 3–19.
© KTK Scientific Publishers, Tokyo, 1985.

GROWTH MECHANISMS AND DEFECTS IN Si LAYERS GROWN ON SiO₂ BY BRIDGING (LATERAL SEEDED) EPITAXY

T. TOKUYAMA, M. TAMURA, N. NATSUAKI, M. OHKURA, M. ICHIKAWA, and M. MIYAO

Central Research Laboratory, Hitachi, Ltd., Kokubunji, Tokyo 185, Japan

Abstract   Mechanisms for lateral seeding growth of Si layers on SiO₂, and defect structures in the regrown layers are discussed based on TEM and micro-probe-RHEED observations. The seeding area essential for lateral epitaxial growth is found to be as small as 0.5 μm wide. It is also found that defects in the regrown layers are related to the time interval during cooling, but are also related to sample structure and laser irradiation conditions. Using 1–2 μm (size and spacing) circular or square seeding patterns, uniform and defect-free regrown layers are developed over a considerably large surface area.

1.   Introduciton

In certain, specific applications of SOI structure where crystal orientation of the grown layer is critical, the use of a laterally seeded growth process generally gives results superior to those obtained using simple growth processes without seeding. [1–4] This process was first reported as lateral Si overgrowth extended to an adjacent SiO₂ layer from the vertically grown epitaxial seeding area during cooling of the laser-melted (pulse) surface Si of a SOI structure. [4]

The first order driving force of such lateral crystal growth originates from the crystallization time delay of melted Si over the SiO₂ layer and the existence of a lateral cooling path through the seeding area, both of which are due to thermal conduction inhomogeneity in the substrate structure. [5,6] Under conventional pulse laser irradiation conditions, however, because such a time delay is rather short, only limited lateral growth has been observed so far, even with a crystallization speed in the order of 100 cm/s. [7]

3

In contrast to stationary pulse laser irradiation, if a movable heating source is used and scanned over the substrate of a speed synchronized to the lateral speed of the crystallization front, considerable longer lateral growth should be possible. In such a case, the selected speed of crystal growth could be between 10^{-3} and 10^3 cm/s (range of growth speeds for Si from the molten phase), however, experimental conditions would limit the speed.

When the substrate temperature is raised by increasing the bias heating, the vertical temperature gradient of the substrate decreases and so-called zone melting at slow growth-speed can be realized by adding a small amount of movable heating. [8] On the other hand, decreasing the bias heating and increasing the movable heating power increases the melting depth of the surface Si layer and raises the temperature beneath the SiO_2 layer, as long as high scanning speed for the movable heating source is not utilized. [9,10]

Speeds of 1 to 10 cm/s have been reported for the latter conditions using a scanning CW laser or an electron beam at a substrate temperature of several hundred degrees. These growth speeds are higher than that known for the CZ or FZ growth of Si crystal ($10^{-3} - 10^{-2}$ cm/s) and are in the order of EFG/ribbon growth. However, Si crystals grown at such speeds have yet to be applied in actual integrated devices.

In addition to the possible effects of growth speed, there are many sources that might cause a defective structure in grown layers ; i.e., thermal stress in the SOI structure, [10] thermal stress originating from melt/solid volume chnge,[11] impurity inclusion in the molten layer from the insulator interface, [12] geometric configuration effect at the edge of the seeding area, [6] overlapping of narrow grown layers, etc. So far, grown crystal layers have only been characterized by measuring the electrical properties of elemental devices fabricated in such layers [13,4,1]; and results show properties somewhat poorer than those of conventional devices fabricated in bulk crystal. Therefore, the results of detailed defect characterization should be presented to optimize growth conditions in order to obtain device-quality crystals.

In this paper, the following subjects are discussed in detail for CW-laser processed, lateral-seeded SOI Si layers using TEM and micro-probe-RHEED[14 - 16] as the diagnosis tools :

   1)   Lateral-seeding epitaxial growth mechanisms.

   2)   Crystal defect observation in grown layers in comparison with those in pulse-laser-processed crystals.

   3)   Effects of geometric configuration and laser scanning methods on growth mechanism.

2. Lateral Seeding Mechanisms

In a sample configuration where poly-Si or amorphous-Si is deposited on SiO₂ film having a window opening that reaches through the Si single crystal substrate, the occurence of epitaxial regrowth of deposited film in this seeding area is essential for observation of lateral regrowth by CW laser irradiation.

In this seeding area, epitaxial regrowth usually occurs between the single crystal substrate and the deposited layer when interface cleanliness and proper laser melting depth are maintained. However, difficulties arise when we try to consider the lateral overgrowth process.

The first point is defect generation in the grown layer of this seeding area. As laser irradiation conditions are usually selected to maintain the particular melting depth most suitable for each sample structure, the heating/cooling thermal gradient in the depth direction becomes considerably large, [17] and this leads to generation of slip-type defects. [18] Also trace of impurities existing at the interface between the substrate and the deposited poly-Si layer play an important role in the generation of defects. A similar situation is also known to exist in pulse laser irradiation. [5,7] Propergation of these defects from the seeding area to the lateral regrowth area depends very much on the thermal conditions that rule lateral growth speed, and should be investigated in detail.

The second point is the differences in irradiation conditions for optimum melting in the seeding area and in layers deposited on SiO₂. This is apparently due to the inhomogeneous nature of thermal conductivity in the substrate structure.

The TEM micrographs in Fig. 1 show after irradiation microstructures for Si layers both on a Si substrate (seeding area) and on SiO₂ (lateral growth area), with laser power as a parameter. The deposited poly-Si layer and SiO₂ thickness were both 350 nm and scanning was with an Ar laser beam having a focus of 60 μm at a speed of 25 cm/s. In the seeding area, irradiation of more than 11W was necessary for vertical epitaxy to occur, whereas in the SiO₂ area, melt-regrowth occured at an irradiation power of less than 10W. It should also be noted in the figure that dense dislocations were generated in the regrown Si in the seeding area.

Other examples of structural change with laser energy are shown in Fig. 2. Here, the laser beam was scanned perpendicular to the SiO₂ stripe patterns. The poly-Si on the SiO₂ was first melted by 8W irradiation (Fig. 2 (a)), and enhancement of grain size was observed in the Si layer over the SiO₂ stripe. Increased grain size on both the SiO₂ and Si substrate (seeding area) can be seen for 9W irradiation energy. It should be noted that grain growth in the poly-Si over the SiO₂ stripe cotinues from the Si substrate even in the case of 8W and 9W irradiation. Continuous single

T. Tokuyama *et al.*

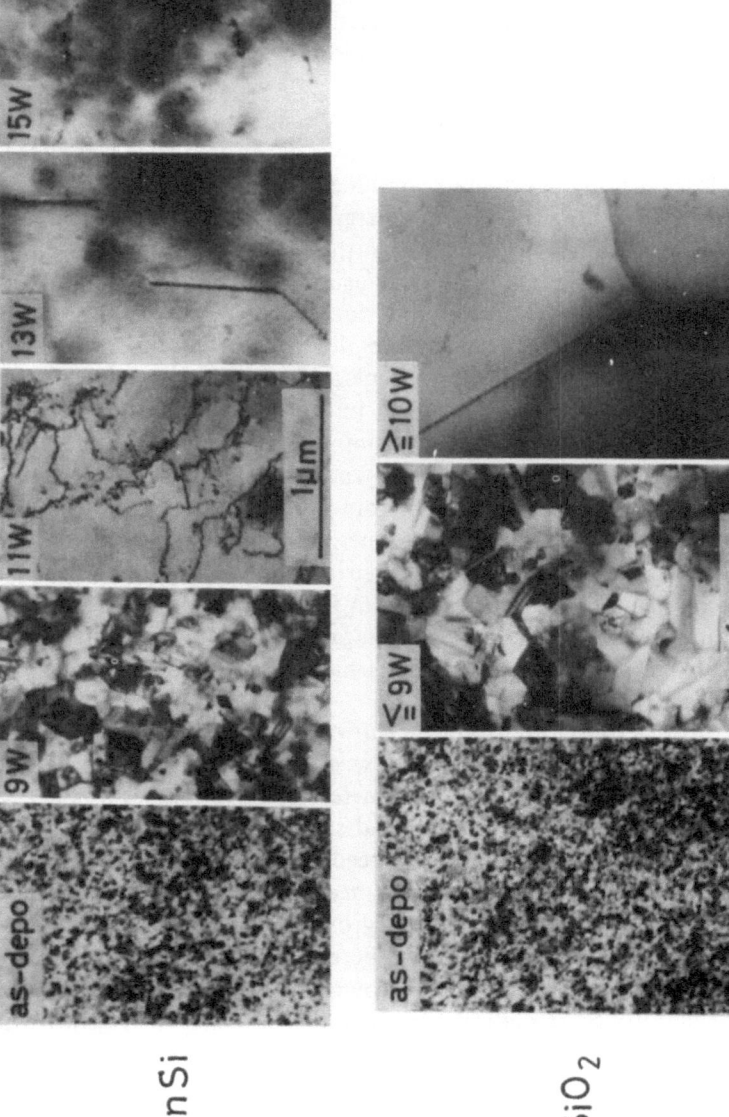

Fig. 1. TEM micrographs of deposited Si film after laser irradiation. CW Ar laser : Scanning speed 25 cm/s, beam diameter 60 μm. Poly-Si film thickness : 350 nm. SiO_2 film thickness : 350 nm (on Si substrate).

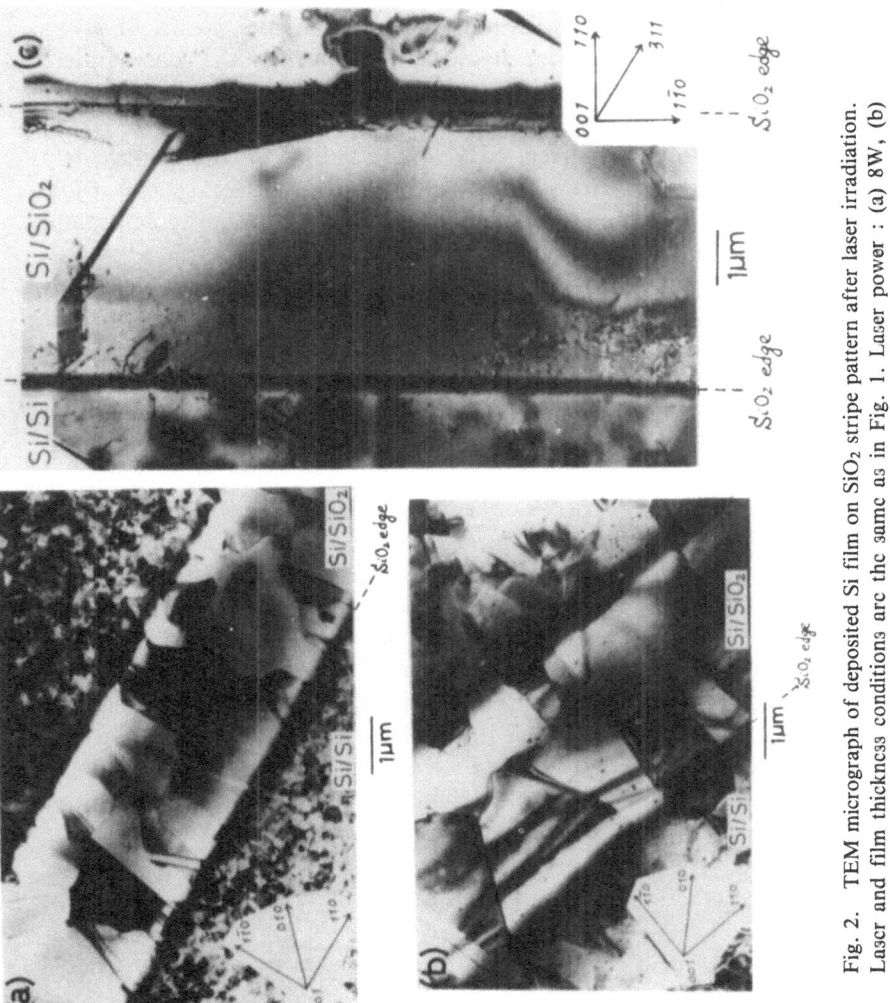

Fig. 2. TEM micrograph of deposited Si film on SiO_2 stripe pattern after laser irradiation. Laser and film thickness conditions are the same as in Fig. 1. Laser power : (a) 8W, (b) 9W, (c) 11W.

crystal film formation was achieved on both Si and SiO₂ stripes with 11
W laser energy.

Suitable thickness anti-reflection film covering each Si layer [19] and
sophisticated built-in thermal structure [20] have both been reported as being effective for prevention of this difference in irradiation conditions.

In order to investigate orientation relation in the grown Si layers,
micro-probe-RHEED measurements were performed. In the general micro-
probe-RHEED system arrangement shown in Fig. 3, the field emmission
fine focus electron beam is incident with the glancing angle to the sample
surface. The absorption-current SEM image shown on the display helps
to determine the sample location to be investigated with a spacial resolution of 0.1 μm. Diffraction spots corresponding to a paritcular sample
location are obtained on the fluorescent screen. [14,15]

Diffraction patterns obtained from various regions of a grown Si layer
are shown in Fig. 4 [21] Single crystal diffraction patterns with Kikuchi
lines were obtained all over the regrown layers. These patterns were all
identical to the patterns obtained from the seeding area ((110) 4° off
the axis), clearly indicating that lateral epitaxy actually did take place during regrowth.

When laser irradiation power was reduced so only regrowth of the
Si layer on SiO₂ would occur, the obtained set of diffraction patterns
similar to the case in Fig.4 becomes as shown in Fig. 5. These show that
the seeding area is still poly-crystal due to imperfect melting of the deposited
layer, whereas a single crystal pattern with Kikuchi lines was obtained
for Si layers on SiO₂. These phenomena had apparently been thought to
be the same as those Magee *et al.* found. [22]

Here again, orientation of the regrown layer is 4° off the (100) axis.
The only possible mechanism that explains these result is that, even though
the deposited layer in the seeding area does not completely melt, in the
vicinity of the SiO₂ layer edge, the temperature rise in the poly-Si is larger

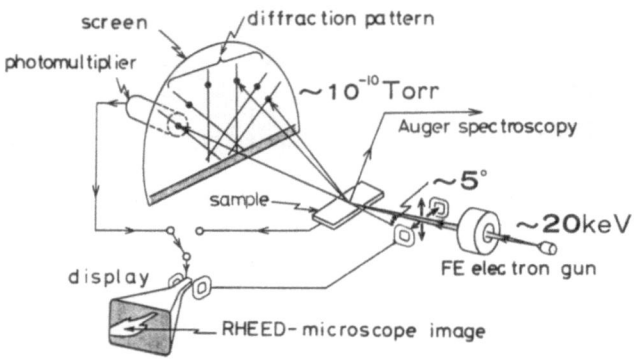

Fig. 3.   Schematic drawing of the micro-probe-RHEED system.

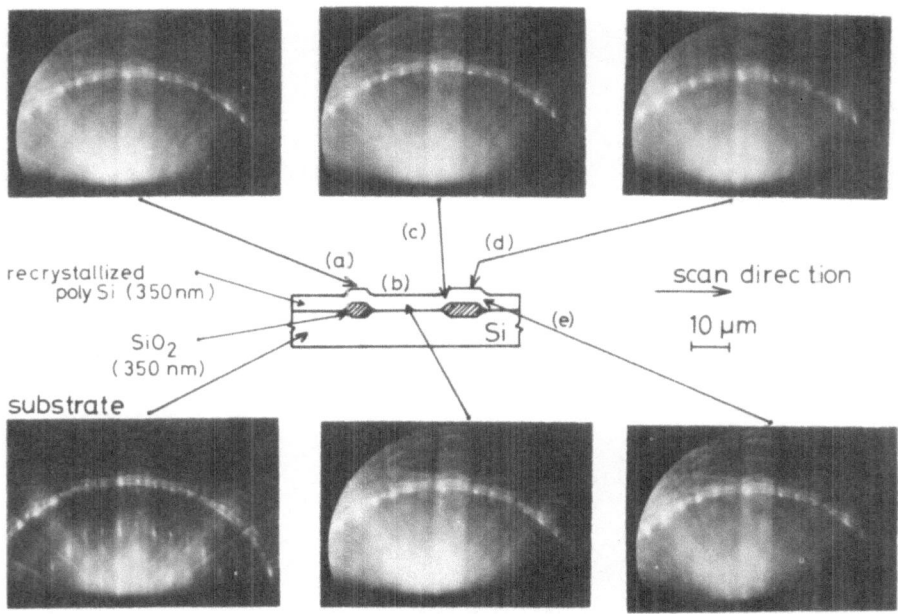

Fig. 4. Diffraction patterns obtained from various regions of a sample irradiated under optimum laser irradiation conditions for lateral growth.

than that in other regions. This is because the wall of the SiO_2 layer prevents heat flow. Thus, in this region, the deposited poly-Si layer melts and recrystallizes into a single crystal by epitaxial growth over the SiO_2 substrate. From the figure, a seeding area even within 0.5 μm of the SiO_2 edge is effective.

Thus, lateral epitaxial overgrowth onto the SiO_2 area from the seeding area is observed in a rather narrow window that depends on laser irradiation conditions. As shown in Fig. 6, for our experimental conditions, optimum growth was only observed in zone II. In zones I and III, peeling of the Si from the SiO_2 (zone I) or incomplete single crystal growth in the seeding area (zone III) was observed.

3.   Defect Structures Observed in Regrown Layers

The detailed TEM micrograph shown in Fig. 7 indicates that there are many (110) oriented slip dislocations in the seeding area. However, these dislocations do not propagate to the regrown Si layers on the SiO_2 as they do in the pulse laser bridging epitaxy case. [7] By increasing the laser irradiation power, the density of these dislocations were observed

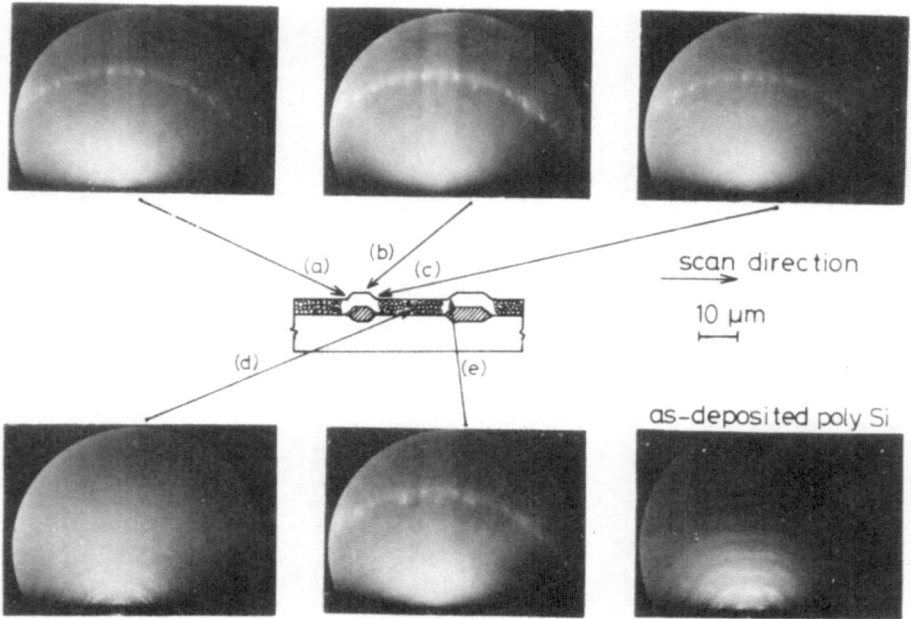

Fig. 5. Diffraction patterns from a sample in which single recrystallization occurred only in the SiO₂ substrate region.

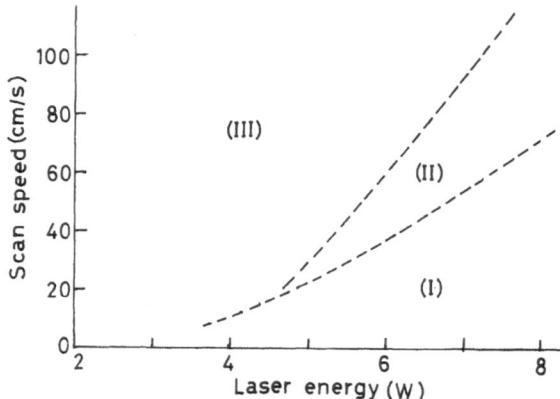

Fig. 6. Laser scanning speed - energy relation for lateral seeding growth. CW Ar laser beam diameter : 20 μm. I : Peeling of Si from SiO₂. II : Opimum condition for lateral epitaxial regrowth. III : No single crystalline Si growth occurs in the seeding area.

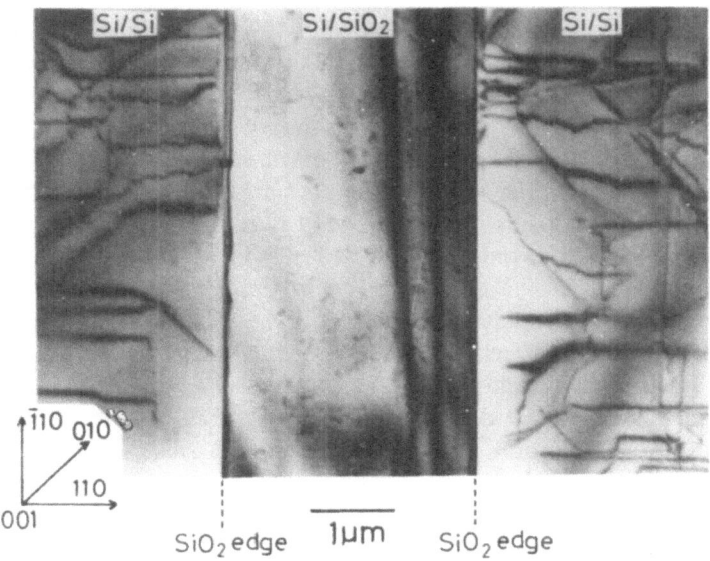

Fig. 7. TEM micrograph showing dislocation networks generated in regrown Si in the seeding area (deposited poly-crystalline seed). Deposited Si film thickness : 350 nm.

to increase. This means that such dislocations are not merely due to thermal stress originating from the Si-SiO₂ structure, but rather they come from the thermal stress originating in the surface localized melting layer. Similar slip dislocations are known to generate in a bulk Si surface during CW laser scanning annealing.

These dislocations were found to be located several microns from the Si surface, [23] where calculations showed a steep temperature gradient. [17]

During lateral growth, at the boundary between the seeding and SiO₂ areas, these dislocations are thought to escape from the surface due to the existence of a step structure in the deposited Si layer. This was also proven by the poor electrical characteristics of fabricated MOS devices that included this particular region as a channel. The effect of dislocations in the seeding area on device characteristics was also revealed by the poor leakage current of a junction diode fabricated in this area.

In pulse laser bridging epitaxial growth, the existence of (311)-oriented line defects was observed in a lateral regrown layer on the SiO₂ area, whereas, as seen in Fig. 7, in the CW laser case, the grown layer is almost defect free. The difference in cooling rates for pulse and CW lasers (10^9 deg/s for pulse and 10^5 deg/s for CW) is assumed to be the cause of this structural difference.

In order to reduce defect generation in the regrown layer in the seeding area, epitaxial (CVD) seed structure was also investigated. In the seeding area, Si single crystal was grown epitaxially by conventional SiH$_4$ thermal decomposition. At the same time, poly-crystalline Si was deposited on the SiO$_2$ area. The single/poly boundary was thus located in the deposited layer corresponding to the Si/SiO$_2$ boundary of the substrate.

Figure 8 shows TEM micrographs of an epitaxial seed SOI structure after laser irradiation. By selecting the proper laser irradiation conditions, stacking faults and twins originally found near the single/poly boundary were eliminated and a continuous single crystalline layer developed all over the sample surface. The defect density in the seeding area was notably smaller than for the poly-seed case.

A scan-rate-power relation similar to that shown in Fig. 6 is plotted in Fig. 9 with typical etch-pit patterns. A slow-scan, high-power condition results in a heavy defect density in the seeding area, whereas a fast-scan, medium-power condition results in a defect-free structure. Thus, in an epi-seed structure, a laser irradiation condition that may not melt the deposited (single crystal) layer on the seeding area but still melts the layer on SiO$_2$ is considered to give favorable results.

4. Effects of the Geometirc Patterns of the Seeding Area

4.1 Effects of laser scanning direction

When the laser is scanned parallel to the SiO$_2$ stripe, and both the width of the SiO$_2$ stripe and the seeding area opening are smaller than the laser beam diameter, Si overgrowth occurs from both sides of the

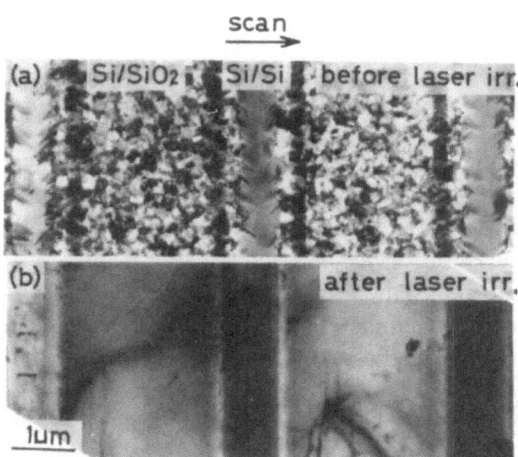

Fig. 8. TEM micrograph of a sample with an epitaxial (single crystal) seeding structure. Deposited Si film thickness : 400 nm.

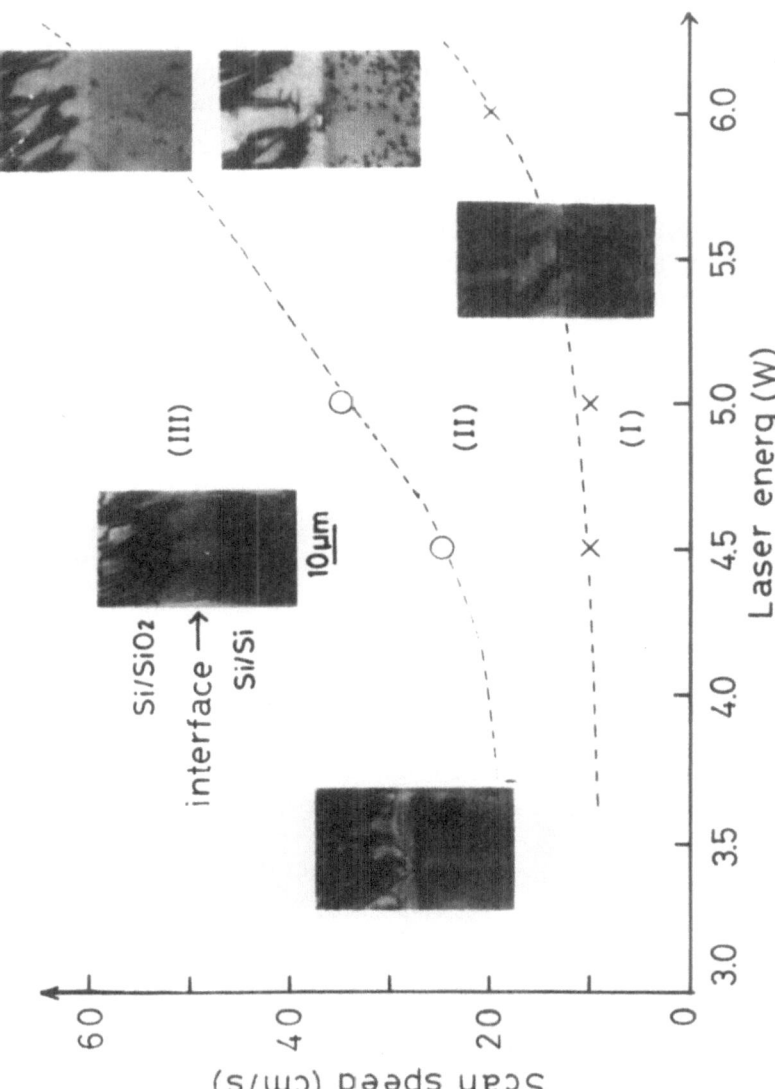

Fig. 9. Laser scanning speed — energy relation for lateral seeding regrowth (epi-seed case). Inserts show Secco etch patterns of samples treated under each laser condition. I : Etch pit density in the seeding area $\gtrsim 2 \times 10^8$ /cm^2. II : $\approx 10^6$ /cm^2. III : $\leq 10^3$ /cm^2

SiO$_2$ edge, and the edges meet at the center of the SiO$_2$ area. As shown on Fig. 10, a weak boundary structure is observed at the center ; however, this boundary structure is not as disticnt as that observed in pulse laser bridging epitaxial growth.

This is because, in scanning parallel to the SiO$_2$ stripe, there is room for molten Si to move in the scanning direction and enough time before cooling occurs for both sides of the single crystals to adjust at the ceter. Obviously, in the case of pulse laser bridging epitaxial growth, these conditions are not met.

4.2 The effects of seeding area patterns

When realistic applications of SOI structures are considered, the sizes and shapes of the SiO$_2$ and seeding area patterns are not necessarily limited to the striped patterns. Figure 11 shows Secco etch micrograph of a regrown surface layer with circular seeding areas (10 μm spacing). In the scanning direction, no defects were found between the seeding areas, whereas in other areas, formation of distinct boundaries was observed. It can be conjuctured that there is considerable stress due to the complicated, inhomogeneous heat flow. Such thermal inhomogeneity can be decreased by selecting a suitable seeding/SiO$_2$ pattern size.

Figure 12 shows TEM micrographs for sample with 1–2 μm circular patterns. In the circular seeding case (Fig. 12 (a)), uniform lateral growth onto the SiO$_2$ area is observed. The rectangular growth boundary surrounding the circular seed area seen in this figure indicates that growth occurred radially, although the skew of this rectangle indicates the existence of laser scanning direction effects.

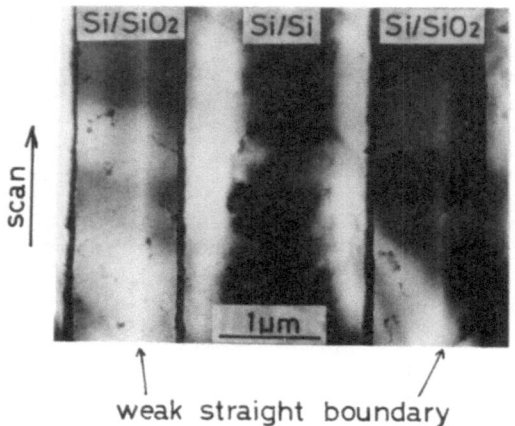

Fig. 10.   TEM micrograph showing lateral regrowth (poly-crystal seed, laser scanning parallel to the SiO$_2$ stripe pattern). Deposited Si film thickness : 350 nm.

scanning direction

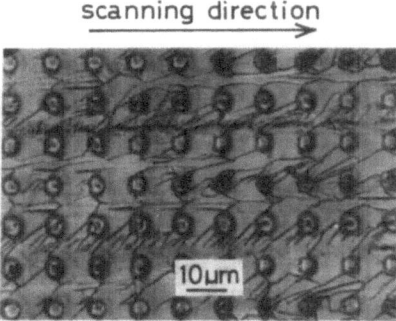

Fig. 11. Secco etch surface micrograph of a regrown layer with circular seeding area. Deposited poly-crystal Si film thickness : 350 nm.

scanning direction

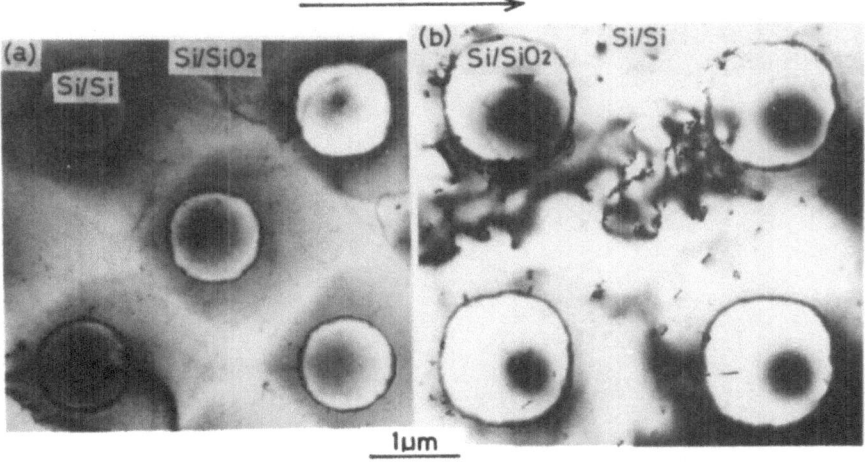

Fig. 12. TEM micrograph of (a) circular seeding area, (b) circular SiO₂ samples after laser irradiation (poly-crystal seed).

In the reverse pattern case (circular SiO_2, Fig. 12 (b)), single crystal regrowth is also observed all over the sample surface. The dislocations observed in the seeding area, and the surface roughness (due to mass flow) found in the laterally grown circular zone are features for this type of sample. Lateral growth circular zone are features for this type of sample. Lateral growth with a small degree of scanning direction effect is also observed in the radial direction onto the SiO_2 circular pattern. This growth in the radial direction is more clearly seen in the reduced power irradiation case shown in Fig. 13.

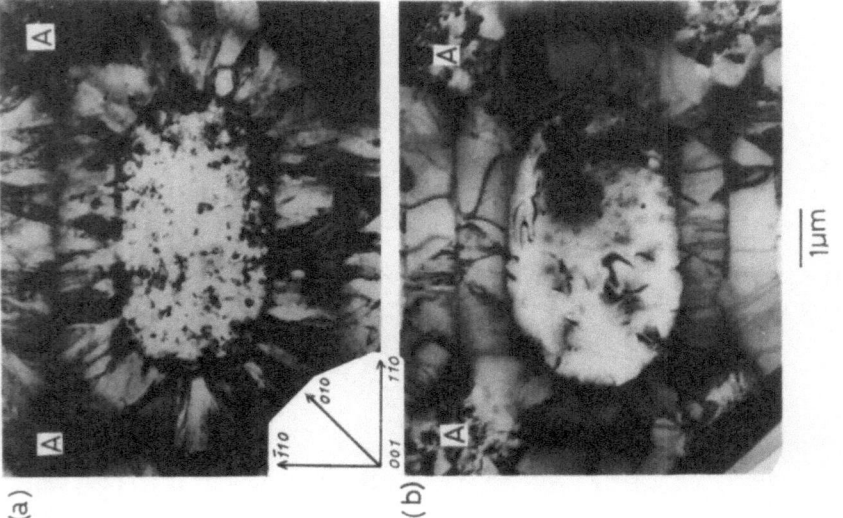

Fig. 14. TEM micrograph of elliptical seeding area samples irradiated by pulse ruby laser. (a) 1.5 J /cm², (b) 1.7 J /cm².

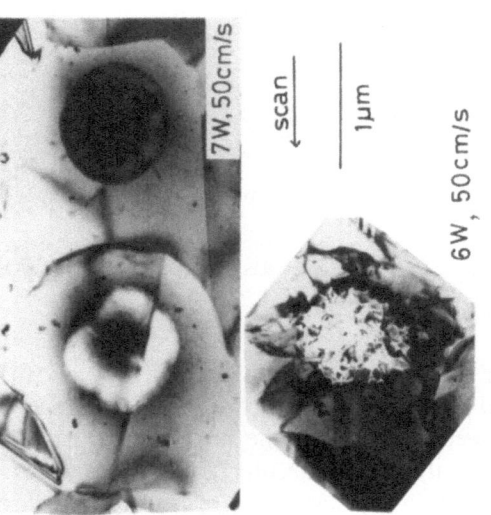

Fig. 13. TEM micrograph of circular seeding area samples after reduced laser energy irradiation.

scan

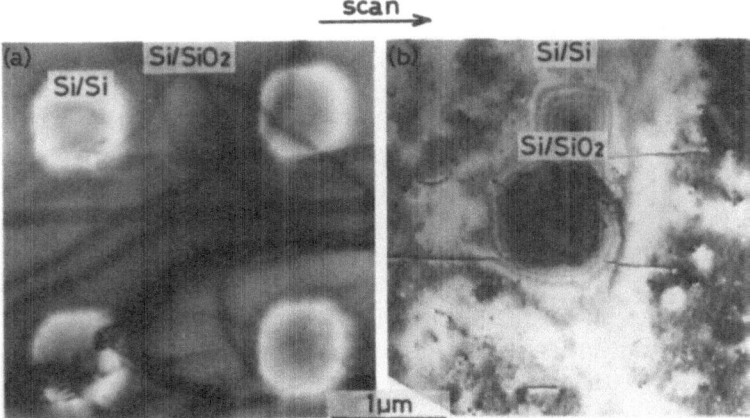

Fig. 15. TEM micrograph of (a) square seeding area, (b) square SiO₂ samples after laser irradiation (epi-seed).

In the pulse irradiation case, however, as shown in Fig. 14, only poly-Si radial regrowth with no preferred orientation is observed. The phenemenological difference between these two growth processes are thought to stem from the existence of (1) Si mass flow during solidification in the laser scanning direction and (2) a time interval long enough to prevent random nucleation during cooling for the CW laser case.

TEM micrographs of epi-seed samples are shown in Fig. 15. The rectangular seed pattern result (Fig. 15 (a)) shows considerable difference in growth compared with that of the poly-seed case (Fig. 12 (a)). The growth boundary seen in the poly-seed sample is not observed here. Instead, many dislocations (some even continuous to the seeding area) are found in the laterally grown area on the SiO₂. Dislocations are also seen in the regrown layer on the small SiO₂ rectangles (Fig. 15 (b)).

Thus, there is considerable difference in the growth process between poly- and epi- seed cases. Since lateral growth follows solidification of the seeding area and it is not necessary for the seeding area to be melted in the epi-seed case, more time for melting of the Si layer over the SiC₂ area before solidification is assumed. The time interval for lateral regrowth is thus shortest for the pulse laser irradiation scheme, medium for the poly-seed and longest for epi-seed CW irradiation scheme. Therefore, in the epi-seed case, dislocations should have sufficient time to propagate before final cooling occurs. The existence of dislocations through the seeding to the lateral growth area found in Fig. 15 (a) indicate this mechanism.

5. Conclusions

Lateral regrowth of single crystal Si layers with an orientation aligned to the substrate were obtained from deposited poly-Si on SiO_2 using various seeding structures and CW laser irradiation. The growth mechanisms and defect structures in the regrown layers were discussed based on TEM and micro-probe-RHEED observation. Precise orientation analysis using micro-probe-RHEED indicated the essential role of seeding area in lateral epitaxial regrowth, although its width can be reduced to as little as 0.5 μm.

TEM analysis of the heavy dislocations found in regrown layers in the seeding area shows that their origin is thermal stress from the steep temperature gradient encountered during cooling. Lateral propagaton, as well as density reduction of these dislocations, were discussed in connection with the sample configurations (size and pattern of the seeding area, single/poly-seed structure, etc.) and laser irradiation conditions that determine the time for cooling.

Using 1–2 μm (size and spacing) circular/square seeding patterns, uniform, defect-free regrown layers were developed over a considerable sample area, which possibly opens the way to sophisticated application to a new integrated device structure.

REFERENCES

1) M. Miyao, M. Ohkura, I. Takemoto, M. Tamura, and T. Tokuyama : Appl. Phys. Lett. **41** (1982) No.1, 59.
2) T. Nishimura, Y. Akasaka, H. Nakata, and A. Ishizu : Appl. Phys. Lett. **42** (1983) No.1, 102.
3) H. W. Lam, R. F. Pinizzotto, S. D. S. Malhi, and B. L. Vaandrager : Appl. Phys. Lett. **41** (1982) No.11, 1083.
4) H. J. Leamy, R. C. Frye, K. K. Ng, G. K. Celler, E. I. Povilonis, and S. M. Sze : Appl. Phys. Lett. **40** (1982) No.7, 598.
5) M. Tamura, H. Tamura, and T. Tokuyama : Jpn. J. Appl. Phys. **19** (1980) L23.
6) G. K. Celler, McD. Robinson, and D. J. Lischner : Appl. Phys. Lett. **42** (1983) No.1, 99.
7) M. Tamura, H. Tamura, M. Miyao, and T. Tokuyama : *Proc. 12th Conf. Solid State Devices, Tokyo, 1980*, Jpn. J. Appl. Phys. Suppl. **20–1** (1981) 43.
8) J. C. C. Fan, M. W. Geis, and B-Y. Tsaur : Appl. Phys. Lett. **38** (1981) No.5, 365.
9) T. I. Kamins, T. R. Cass, C. J. Dell'Oca, K. F. Lee, R. F. W. Pease, and J. F. Gibbons : J. Electrochem. Soc. **128** (1981) 1151.
10) H. W. Lam, R. F. Pinizzotto, and Al F. Tasch, Jr. : J. Electrochem. Soc. **128** (1981) 1981.
11) R. F. Pinizzotto, H. W. Lam, and B. L. Vaandrager : Appl. Phys. Lett. **40** (1982) No.5, 388.
12) H. J. Lamy : *Laser and Electron-Beam Interactions with Solids*, ed. B. R. Appelton and G. K. Celler (North-Holland, New York, 1982) p. 459.
13) B-Y. Tsaur, M. W. Geis, J. C. C. Fan, D. J. Silversmith, and R. W. Mountain : Appl. Phys. Lett. **39** (1981) No.11, 909.
14) M. Ichikawa and K. Hayakawa : Jpn. J. Appl. Phys. **21** (1982) 145.
15) M. Ichikawa and K. Hayakawa : Jpn. J. Appl. Phys. **21** (1982) 154.

16) M. Ichikawa, M. Ohkura, and K. Hayakawa : Jpn. J. Appl. Phys. **22** (1983) 527.

17) I. D. Calder and R. Sue : J. Appl. Phys. **53** (1982) 7545.

18) M. Tamura, M. Ohkura, and T. Tokuyama : *Proc. 13th Conf. Solid State Devices, Tokyo, 1981*, Jpn. J. Appl. Phys. Suppl. **21-1** (1982) 193.

19) J. P. Colinge, E. Demoulin, D. Bensahel, and G. Auvert : *Proc. 14th Conf (1982 International) Solid State Devices, Tokyo, 1982*, Jpn. J. Appl. Phys. Supp. **22-1** (1983) 205.

20) S. Kawamura, N. Sasaki, T. Iwai. M. Nakano, and M. Takagi : This volume, pp. 67-84.

21) M. Ohkura, M. Ichikawa, M. Miyao, and T. Tokuyama : Appl. Phys. Lett. **41** (1982) No.11, 1089.

22) T. J. Magee, L. J. Palkuti. R. Ormond, C. Leung, and S. Graham : Appl. Phys. Lett. **38** (1981) No.4, 248.

23) S. Minagawa, K. F. Lee, J. F. Gibbons, T. J. Magee, and R. Ormond : J. Electrochem. Soc. **128** (1981) 848.

24) M. Miyao, M. Ohkura, T. Warabisako, and T. Tokuyama : *Laser-Solid Interactions and Transient Thermal Processing of Materials* ed. J. Narayan, W. L. Brown, and R. A. Lemons (North-Holland, New York, 1983) p. 499.

Silicon-on-Insulator: Its Technology and Applications, edited by S. Furukawa, pp. 21–28.
© KTK Scientific Publishers, Tokyo, 1985.

LASER CRYSTALLIZATION OF POLYCRYSTALLINE SILICON BY CONTROLLING LATERAL THERMAL PROFILE

T. Nishimura,[1] Y. Akasaka,[1] H. Nakata,[1]
K. Sugahara,[2] and T. Isu[2]

[1]*LSI Research and Development Laboratory, Mitsubishi Electric Corporation, Itami 664, Japan*
[2]*Central Research Laboratory, Mitsubishi Electric Corporation, Amagasaki 661, Japan*

Abstract The thickness variation of the silicon dioxide layer underlying polycrystalline silicon (polysilicon) causes the effective thermal profile in the polysilicon film for controlling the nucleation process during laser recrystallization. The obtainable grains are rectangular in shape as large as 8 μm × 5C0 μm.

Raman microprobe is used to measure local strain at various points. The observation indicates that the residual tensile stress in the recrystallized silicon films on the structured oxide layer in this study is about 1/3 as small as that in laser-recrystallized polysilicon films on a flat oxide layer.

1. Introduction

Recently, a considerable amount of work has been devoted to achieving the high quality silicon film on an insulating layer (SOI) by laser recrystallization technique. Among those studies, prime concerns were in the control of the nucleation process to enlarge the grain size of as-deposited polycrystalline silicon (polysilicon) without any seed.

Several attempts have been made to control the nucleation process, by controlling the power absorption in substrate,[1] by shaping the laser beam,[2,3] by varying the power absorption with the patterned antireflecting coating,[4] or by using the fast sinusoidal x- and slow linear y-scan,[5] as shown in Fig. 1. Basically all methods intended to produce the concave melt-resolidification front in the silicon film at recrystallization process.

In this study, we tried to control the heat flow in substrate with thickness variation of the silicon dioxide layer (oxide layer) underlying

CONTROL OF THERMAL PROFILE

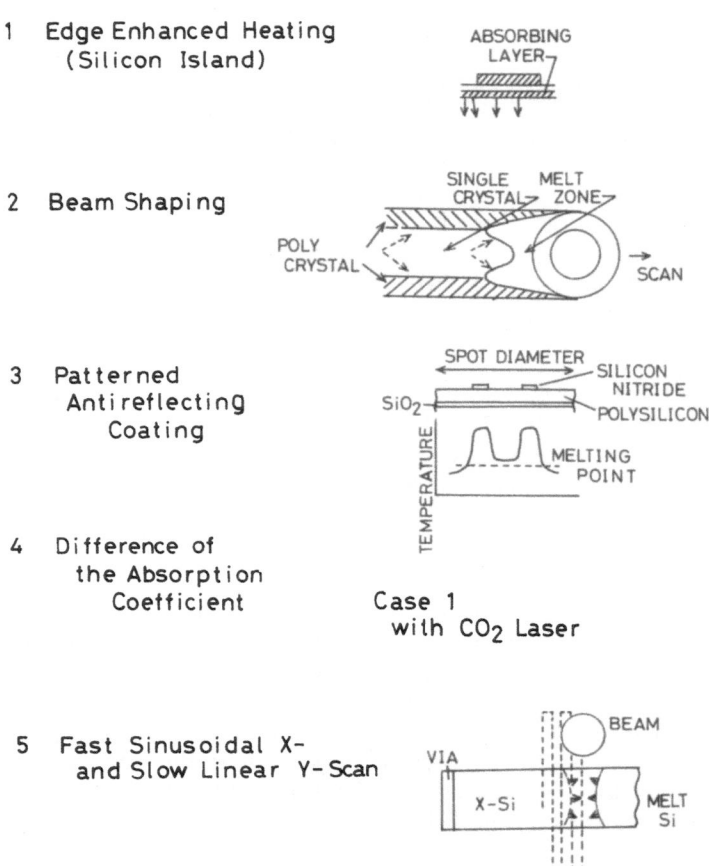

1   Edge Enhanced Heating
    (Silicon Island)

2   Beam Shaping

3   Patterned
    Antireflecting
    Coating

4   Difference of
    the Absorption
    Coefficient

5   Fast Sinusoidal X-
    and Slow Linear Y-Scan

Fig. 1.   Methods for controlling the nucleation process.

polysilicon, and obtained a large grain growth. For evaluation of the crystal quality, Raman microprobe with spatial resolution of 1 μm was applied. It was found that the residual stress in recrystallized polysilicon was about one third as small as those in laser recrystallized silicon films on flat insulator.[6,7]

2. *Experiments*

The starting sample was an [100] 4-inch silicon wafer with an 1 μm thick thermally oxidized SiO_2 layer. A surface relief grating was defined and etched into the oxide layer by the reactive ion etching technique. The

width and height of grating relief were 2 μm and 0.18 μm, respectively.
The space between the parallel stripes was varied from 2 μm to 3 μm.
Then, a 7000 Å thick low pressure chemical vapor deposition (LPCVD)
polysilicon was deposited on the oxide layer, and capped with a 650 Å
thick silicon nitride film. The schematic cross section of the sample is
shown in Fig. 2. The sample was scanned with a cw argon laser at a
scan speed of 12.5 cm/sec, spot size of 40 μm, laser power of 3–4 W,
and substrate temperature of 450°C. The scan direction was parallel to
the grating stripes.

3. Experimental Results and Discussions

3.1 Control of the lateral thermal profile with the thickness variation of the oxide layer underlying polysilicon

As the thermal conductivity of SiO_2 is very small, the resultant lateral
variation of heat flow in substrate is considered to be small. However,
by adjusting the dimension of the structure, the distribution of heat flow
in substrate depending on the thickness variation of the oxide layer underlying polysilicon is expected to produce the temperature gradients in
recrystallized polysilicon film. If it becomes concave, the solidification
from melting will proceed to grow larger grains using the already recrystallized large grain area as a seed.

Figure 3 shows photomicrographs of the recrystallized polysilicon films
whose grain boundaries were delineated with the secco etching. At proper
laser power, large grains grew between the grating stripes. Obtainable grains
were rectangular in shape as large as 8 μm × 500 μm. In order to evaluate

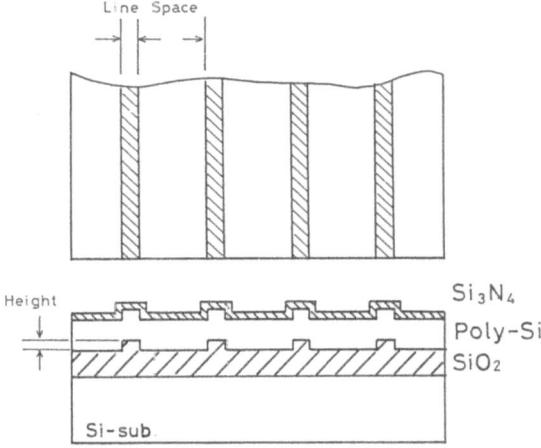

Fig. 2. Schematic diagram of the sample structure.

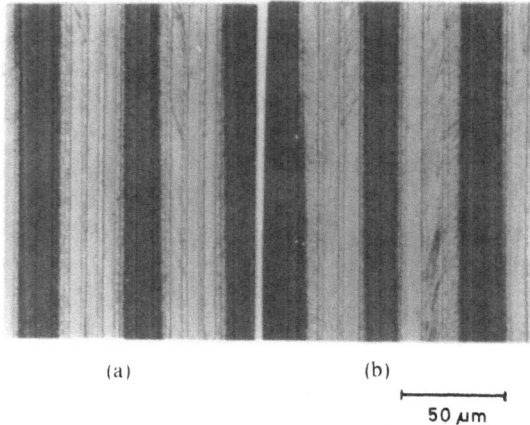

(a)   (b)

50 µm

Fig. 3. Photomicrographs of laser-recrystallized polysilicon films on a grating structured insulating layer, a) grating space of 6 µm, b) grating space of 8 µm.

the nucleation process in more detail, the sample was prepared for the transmission electron microscope (TEM) and observed as shown in Fig. 4. The grain boundary was located on the thick oxide stripes along with their orientation, and/or ceased extending from the surrounding fine grains onto the regions. This indicates that the area on the thick oxide stripes finally cooled down, continuing the crystal growth. So it is confirmed that the temperature rise in the polysilicon on those stripes was higher than that in the polysilicon layer in between the grating stripes.

The certainty for the larger grain growth and its reproducibility was

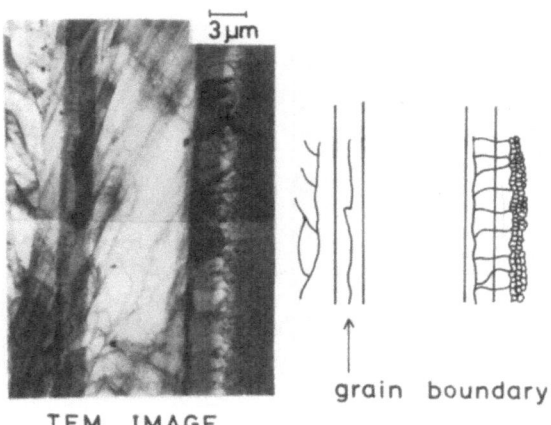

3 µm

grain boundary

TEM IMAGE

Fig. 4. Transmission electron microscopic view of the laser-recrystallized polysilicon on a structured oxide layer with a grating space of 8 µm.

increased as the grating height was increased. This is naturally because the larger temperature gradient is produced by the larger heat flow in substrate through the thinner oxide layer. In addition, the certainty was also increased by increasing the thickness of the silicon nitride layer which was deposited for encapsulating. In this case, though the mechanism has not been understood clearly, it is qualitatively considered that the silicon nitride layer contributes to make homoginize the gaussian like power distribution of the laser beam, since thermal diffusivity of the silicon nitride is not so small as compared with that of silicon.

Several crystallographic orientations were obtained from enlarged grains by transmission electron diffraction (TED) measurement. The major orientations were [100] and [110], but [111], [211] and [310] were also observed. Table 1 shows the summary of the observed orientations.

3.2 Evaluation by Raman microprobe

The Raman microprobe[6] was applied to evaluate the residual stress in the recrystallized polysilicon film with a spatial resolution of 1 μm. The experimental apparatus was basically the same as that used in ref. 6. The 488.0 nm line of an argon laser was used for incident beam. The scattered light was collected by an objective lens, and directed into a double monochrometer. Detected signals were then transformed to the Raman spectra.

Figures 5-a and 5-b show the photomicrographs of recrystallized polysilicon films on this sample structure and the flat oxide layer, respectively. The schematic cross sections are also shown. In the case of the flat oxide layer, the well known chevron like feature is seen. The circles indicate the position of the laser spot used in the Raman scattering measurements. Figure 6 summarizes the frequency of Raman band and the full width at half-maximum (FWHM) for various positions, and for single crystalline silicon and as-deposited polysilicon. The frequency of the Raman band from the recrystallized polysilicon film on the flat oxide layer was about 518 cm^{-1}, which was 3 cm^{-1} lower than that in unstressed silicon crystals. This peak shift to lower energy is considered to be

Table 1.   Summary of the observed crystallographic orientations.

Orientation	Appearances
⟨100⟩	3/10
⟨110⟩	3/10
⟨111⟩	1/10
⟨210⟩	1/10
⟨310⟩	1/10
unknown	1/10

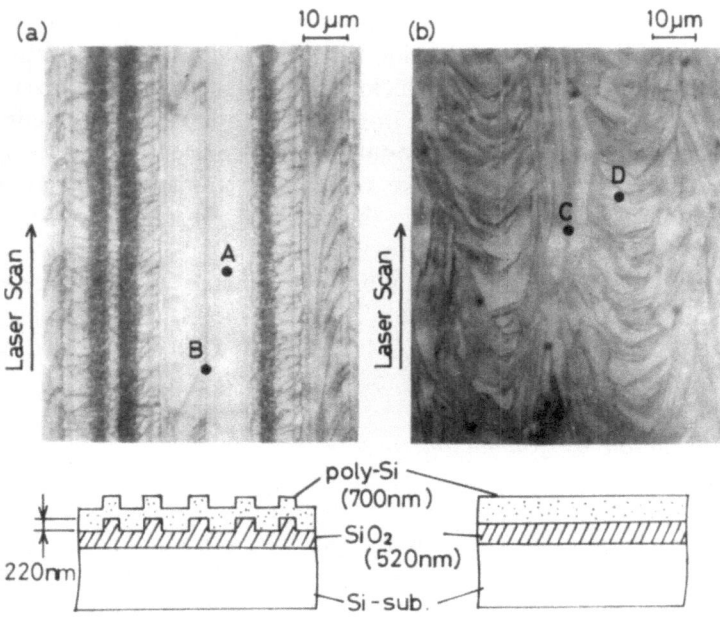

Fig. 5. Photomicrographs of laser-recrystallized polysilicon a) on a structured oxide layer, b) on a flat oxide layer, and their schematic cross section.

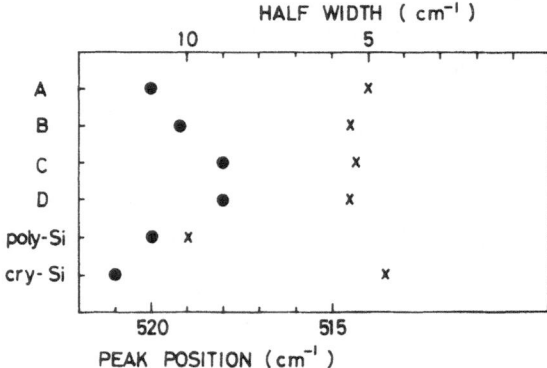

Fig. 6. The measured frequency of the Raman band and full width at half maximum (FWHM) for various points and for single crystalline silicon and as-deposited polysilicon. • The frequency of the Roman band. × FWHM.

caused by the uniform tensile stress, which is mainly due to interfacial pinning at solidification temperature followed by differential contraction.[6-8] Although measurements were done at only two points in this experiments, it had been confirmed that the magnitude of the strain was independent of position.[7]

In the case of the structured sample, the Raman peaks from the enlarged grain and the vicinity of grain boundary were about 520 cm^{-1} and 519 cm^{-1}, respectively. Especially in the case of the enlarged grain, the resulting peak shift was less than 1 cm^{-1}. Since any splitting in Raman spectra was not observed, it was also assumed that the observed peak shift was due to the hydrostatic component of the strain, not the uniaxial stress. Thus the residual tensile stress was estimated to be $1-3 \times 10^9$ dyne/cm^2 which was about 1/3 as small as that on a flat oxide layer. The residual stress in the vicinity of the grain boundary was slightly large as compared with the value from the center of the enlarged grain, but was still smaller than those in recrystallized polysilicon on a flat oxide layer. This small residual stress implies the existence of some relaxation mechanism or modification of nucleation process that might be related to the steep curvature of the melting front, the resulting spatial nonuniformity in expansion of molten silicon at solidification.

As a present step, a quantitative discussion is difficult, and requires more experiments. For detail analysis with Raman spectra, it would be required to take into account the polarization of incident and scattered light adjusting the crystal orientation.

4. Conclusion

The thickness variation of the oxide layer underlying polysilicon could produce the effective thermal profile in the laser-crystallized polysilicon film for controlling the nucleation process, and the resulting enlarged grains were rectangular in shape as large as 8 μm $\times$ 500 μm.

By using the Raman microprobe for measuring the local strain in the laser-recrystallized polysilicon film, it was found that the residual stress in the polysilicon film on the structured oxide layer was about 1/3 as small as those in laser-recrystallized polysilicon films on a flat oxide layer.

Acknowledgements
The authors are grateful to Dr. S. Nakashima for his helpful discussion, and to Dr. H. Oka for his interest and support of this research program.
This work was performed under the management of the R&D Association for Future Electron Devices as a part of the R&D Project of Basic Technology for Future Industries sponsored by Agency of Industrial Science and Technology, MITI.

REFERENCES

1) D. K. Biegelsen, N. M. Johnson, D. J. Bartelink, and M. D. Moyer: Appl. Phys. Lett. **38** (1981) 150.
2) S. Kawamura, J. Sakurai, M. Nakano, and M. Takagi: Appl. Phys. Lett. **40** (1982) 394.
3) T. J. Stultz, and J. F. Gibbons: Appl. Phys. Lett. **39** (1981) 498.
4) J. P. Colinge, E. Demoulin, D. Bensahel, and G. Anvert: Appl. Phys. Lett. **41** (1982) 346.
5) G. K. Celler, L. E. Trimble, K. K. Ng, H. J. Leamy, and H. Baumgart: Appl. Phys. Lett. **40** (1982) 1043.
6) S. Nakashima, Y. Inoue, M. Miyauchi, A. Mitsuishi, T. Nishimura, T. Fukumoto, and Y. Akasaka: Appl. Phys. Lett. **41** (1982) 524.
7) S. Nakashima, S. Oima, A. Mitsuishi, T. Nishimura, T. Fukumoto, and Y. Akasaka: Solid State Commun. **40** (1981) 765.
8) K. Kugimiya, G. Fuse, and K. Inoue: Jpn. J. Appl. Phys. **21** (1982) L19.

Silicon-on-Insulator: Its Technology and Applications, edited by S. Furukawa, pp. 29–39.
© KTK Scientific Publishers, Tokyo, 1985.

ELECTRON BEAM RECRYSTALLIZED SOI STRUCTURES

K. Shibata, T. Inoue, K. Kato, and M. Kashiwagi

*Toshiba Research and Development Center, Toshiba Corporation, I,
Komukai-Toshiba-cho, Saiwai-ku, Kawasaki 210, Japan*

Abstract Electron beam recrystallization of silicon films on silicon dioxide has
been studied in three principal subjects. The first, with coalescence of dislocations,
grain boundaries were generated at strain concentrated regions in the recrystallized
silicon film. The second, measuring the transient temperature profile, it was confirmed that when the electron beam scanning rate was above 5 mm/s, the
temperature profile has a convex tailing. The third, ultra high vacuum deposited
silicon films were used for laterally seeded epitaxial growth on silicon dioxide.
The maimum epitaxial growth length and silicon film quality were improved in
comparison with the results of chemically vapor deposited polycrystalline silicon
films.

1. Introduction

Recrystallization of silicon films on insulating substrate (SOI) to single
crystalline has received much interest, because it is a promising technique
to realize multilevel stacked very large scale integrated circuits. The SOI
has also potential in alternation to silicon films on sapphire. There have
been numerous reports on developing the SOI technique, and various
methods have been used to recrystallize the silicon films. In these methods,
electron beam recrystallization is a very interesting one because of its several
advantages.[1] Using the electron beam recrystallization, the significant increase in grain size has been realized. Since grain boundaries affect the
electronic properties of SOI, further efforts are needed to increase the
average grain size.

In this paper, three principal subjects in electron beam recrystallization of SOI were studied. The first part of this paper describes the observation of the grain boundary generation at the stress concentrated region
in recrystallized silicon films. In the second part, transient temperature
profiles during electron beam recrystallization has been measured by a
thermovision. Laterally seeded epitaxial growth of ultra high vacuum (UHV)

deposited silicon films on SiO_2 are demonstrated in the third part.[2]. Both spot and pseudo-line shaped electron beams were used in the recrystallization of silicon films.

2. Grain Boundary Generation

In SOI recrystallization by electron or laser beam, solidification occurs at the tailing edge of the liquid-solid interface. Therefore, the recrystallized grain structure strongly depends on the shape of the liquid-solid interface. Using a crescent-[3] or a doughnut-shaped[4] beam, the long and wide single crystalline silicon area is obtained at the center of the beam scanning region. In these cases, the liquid-solid interface is concave with respect to the solidification front, and individual grains near the center tend to propagate toward the outer edges of the beam scanning region, resulting in wider grains.

In this experiment, the pseudo-line electron beam was mainly used for recrystallization of SOI. This pseudo-line beam was made by high frequency reflection of a spot beam in the rectangular direction to the beam scanning direction.[5] The frequency of reflection was up to 1 MHz, and 300 μm long line beam was obtained. The pseudo-line beam can easily control the liquid-solid interface with oscillation waveform, frequency, beam scanning rate, and substrate temperature.

Figure 1 shows a typical result of the concave liquid-solid interface with respect to the solidification front. This is a Nomarski interference micrograph of the recrystallized silicon surface delineatd the grain boundaries. It can be seen from this micrograph that the grains near the center

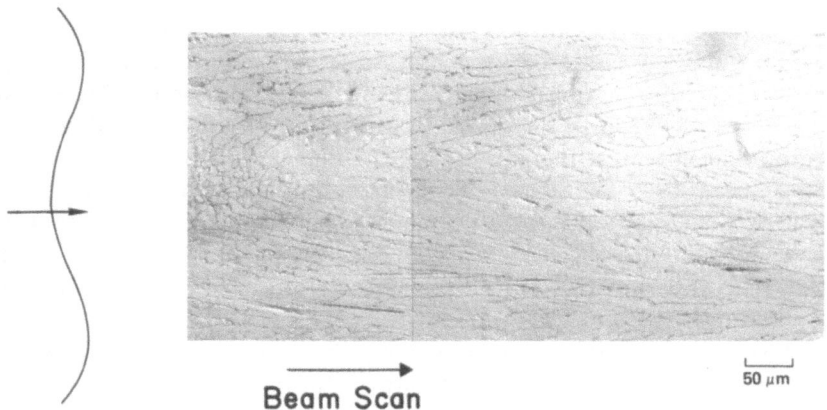

Beam Scan

50 μm

Fig. 1. Pseudo-line electron beam recrystallized silicon film surface. The grains near the center propagate to outer edges of the beam scanning region.

propagate to outer edges of the beam scanning. However, numerous grain boundaries are generated in the middle of the recrystallized region one after another.

In order to study grain boundary generation, a closer investigation has been carried out. Figure 2 shows the grain boundaries at the edge of the beam scanning of the silicon films recrystallized by a spot beam. In this photograph, A is out of the beam scanning region, and silicon film is not recrystallized. B is the transition region between fine grained and large grained regions, whose grain size is submicrons. In C region, individual grains grow inward to the beam scanning center, resulting in large grained polycrystalline silicon. However, in D region, numerous grain boundaries are generated, and large grains are broken up into smaller grains.

To clarify the detailed grain structure and grain boundary generation, the recrystallized silicon films were studied by transmission electron microscope (TEM). A typical result of TEM observation is shown in Fig. 3 (A). It is notable that severe residual strain concentrates at the center of a large stretched grain. An advanced observation at the strain concentrated region was carried out, which is shown in Fig. 3 (B) with higher magnification. It is confirmed that high density dislocations are generated at the strain concentrated region, and that several grain boundaries are generated with the coalescence of these dislocations. The mechanism of the strain concentration in the recrystallization process has not been clearly defined. However, the strain in the recrystallized silicon film would be induced by the differential thermal contraction between silicon and silicon dioxide, or volume expansion of freezing silicon.

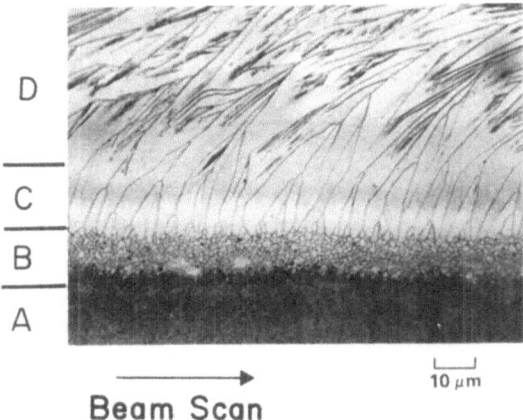

Fig. 2. Grain boundaries at the edge of the electron beam scanning. In D region, numerous grain boundaries are generated, and large grains are broken up into smaller grains.

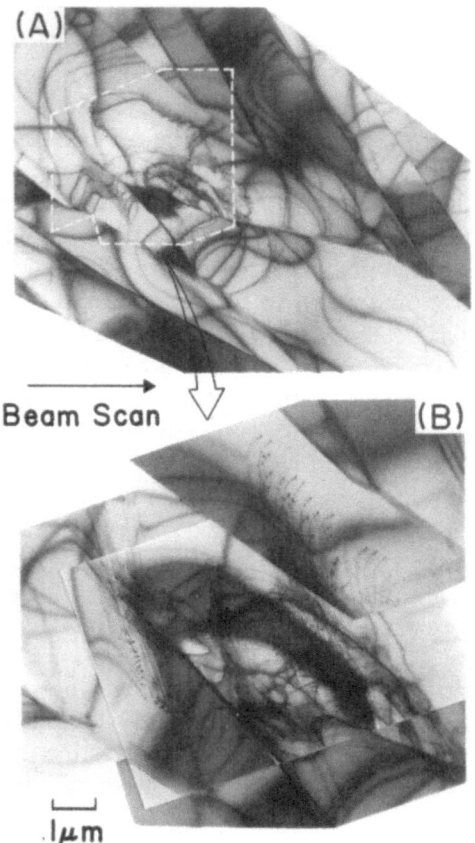

Beam Scan

Fig. 3. Transmission electron micrographs which show (A) grain boundary generation at the strain concentrated region, and (B) high density dislocations and grain boundaries with the coalescence of the dislocations.

3. Transient Temperature Profile Measurement

In the previous section, we described the grain boundary generation at the strain concentrated region in the recrystallized silicon films. The complete elimination of grain boundaries is necessary to realize high quality materials for device fabrication. For this purpose, it is necessary to understand the recrystallization process of thin silicon films, which requires a knowledge of the transient temperature profile during the recrystallization.[6,7]

In this section, we present the measurement of the temperature profile of silicon films during the electron beam recrystallization. The 0.6

μm thick chemically vapor deposited (CVD) polycrystalline silicon films with 0.2 μm thick SiO_2 layer encapsulation were deposited on 1 μm thick thermal SiO_2 layer. Electron beam recrystallization was carried out with an accelerating voltage of 10 kV, typical beam current of 1 to 2 mA, and substrate heating of 500°C.

The transient temperature profile of silicon films was measured by a thermovision, which utilizes an InSb infrared photodetector. Black body radiation images from the wafer were taken through a sapphire viewing port of the electron beam annealing chamber, and displayed on the video monitor.

Figure 4 shows a typical three-dimensional black body radiation image. X- and Y- axes show the wafer plane, and Z-axis shows signal intensity of the black body radiation. A round circle in the flat relief indicates the wafer region, and a steep gaussian-like peak at the midst in the round circle shows emphasized temperature profile of electron beam irradiating region.

In order to obtain the detailed temperature profile, the cross sectional profile of the hot region is measured. Figure 5 shows typical cross sectional profiles at the beam scanning rate of 1 mm/s, and 5 mm/s. Electron beam recrystallization was carried out by the pseudo-line beam. In photograph A, the electron beam scanning rate is 1 mm/s, and the black body radiation intensity profile is symmetric. In photograph B, however, there is a convex tailing on the left shoulder of the profile. This result indicates that the beam scanning rate has a great influence on the temperature profile and the solid-liquid interface shape. So the beam scanning rate is an important parameter in SOI recrystallization.

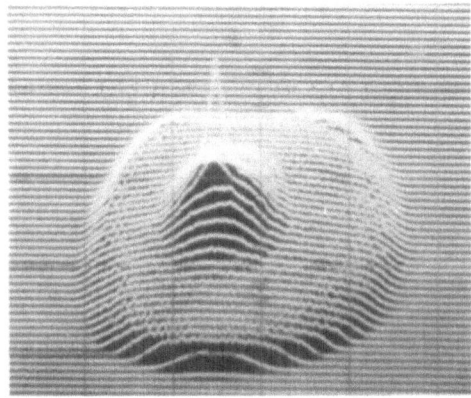

Fig. 4. Typical three-dimensional black body image displayed by the thermovision. A round circle shows the wafer region, and a steep gaussian -like peak shows the electron beam irradiating region.

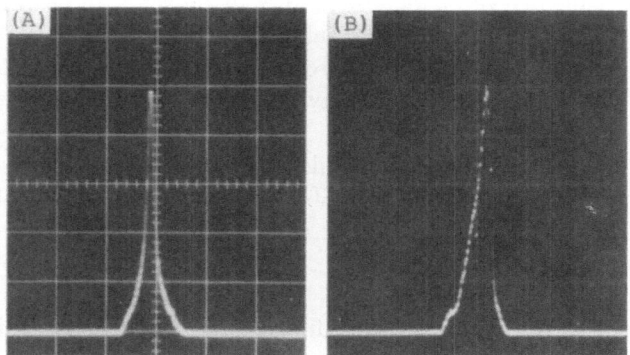

Fig. 5. Typical cross sectional profiles at the beam scanning rate of (A) 1 mm/s, and (B) 5 mm/s. In photograph (B), there is a convex tailing on the left shoulder of the profile.

4. Laterally Seeded Epitaxial Growth

Another common problem which presents itself in the SOI technique is how to control the crystal orientation of recrystallized silicon film. Laterally seeded epitaxial growth has been believed to be advantageous in controlling the surface orientation of SOI film, and in determinating the lithographical position of active devices in SOI.

In most of reported experiments, CVD polycrystalline silicon films with pyrolysis of silane were used for laterally seeded epitaxy.[8] In CVD polycrystalline silicon films, however, impurities such as hydrogen, oxygen, and nitrogen are introduced during the film deposition process. These impurities has been considered to segregate at defects and grain boundaries, and to prevent the successfully seeded epitaxial growth.

The experiments reported that both vertical and lateral epitaxy were needed for growth of single crystalline silicon films on silicon dioxide layer, as shown in Fig. 6(A). In this situation, it is necessary to melt both silicon layer on the seeded area and on SiO_2, adequately. Therefore,'the laterally seeded epitaxy depends on heat disspation difference between the seeded and SOI areas.

In our experiment, both silicon film deposition and electron beam rcrystallization were carried out under UHV condition. Silicon wafers of (100) orientation with 0.2 μm thick thermal SiO_2 layer were chemically etched, resulting in seeded areas of 6 μm wide stripe spacings, and SOI areas of 6 to 20 μm wide stripe lines. Before the silicon film deposition, the wafers were cleaned by rinsing in an HCl, H_2O_2, and H_2O mixed solution, and by heating up to 800°C in an UHV chamber of 2×10^{-10}

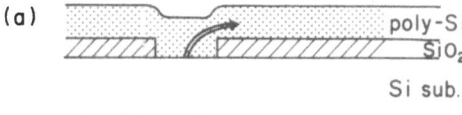

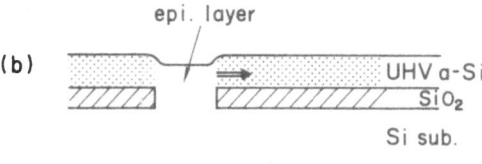

Fig. 6. Schematic diagrams of laterally seeded epitaxial growth, using (A) CVD polycrystalline silicon film, and (B) UHV deposited silicon film.

Torr. Silicon films of 0.5 to 2.0 μm thickness were evaporated from E-gun source with deposition rate of 3 nm/min at substrate temperature of 370 to 600°C.

At substrate temperature above 370°C, silicon layer on the seeded area has epitaxially grown during the deposition process, as shown in Fig. 6(B), which was confirmed by RHEED pattern analysis. This structure does not need complete melting of the silicon layer in the seeded area. The margin of laterally seeded epitaxy was expanded by using the UHV deposited silicon films.

Electron beam recrystallization was subsequently carrid out with a scanning electron beam accelerated at 10 kV. Beam current was typically 2.5 to 3.5 mA, and beam scanning rate was 50 to 100 mm/s.

Figure 7 shows a typical result of Nomarski interference microscope observation of the electron beam recrystallized sample surface. To delineate grain boundaries, the sample surface was chemically etched in mixed acid solution. However, no grain boundaries were found on the surface, and the lateral epitaxy was completely accomplished on 20 μm line width SOI patterns.

Figure 8 shows Rutherford backscattering and channeling spectra of 350 keV protons. About 77 percent of the analyzed area corresponds to the laterally seeded silicon layer on SiO_2. Normalized minimum yield is 0.24, and uniformly distributes throughout the SOI film. Comparing with the (100) bulk silicon result, there exists high density crystalline defects, but it is comparable to that of SOS films at the silicon sapphire interface.

Figure 9 shows electron channeling patterns of the laterally seeded

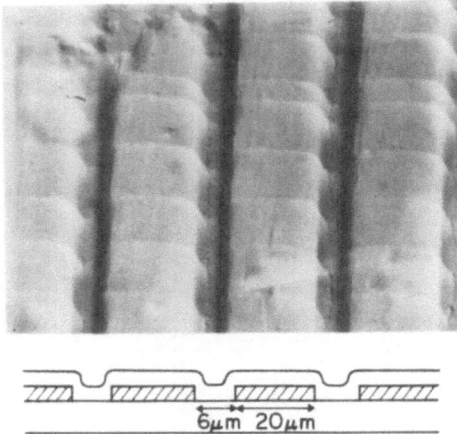

Fig. 7. Typical result of Nomarski interference microscope of the complete epitaxial growth sample.

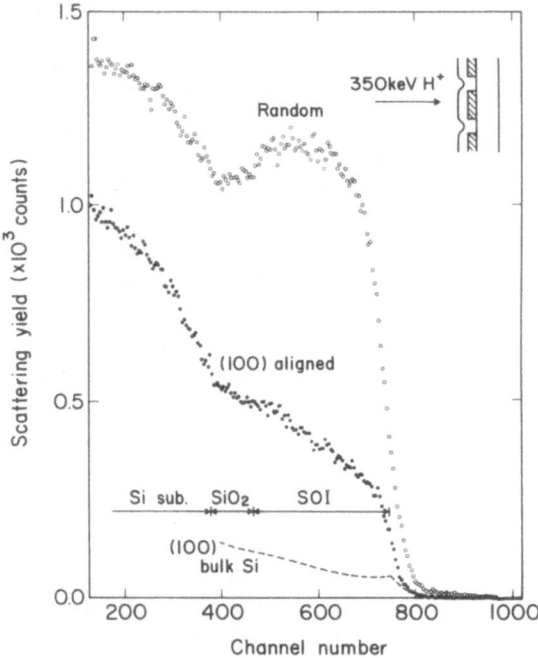

Fig. 8. Rutherford backscattering and channeling spectra from the electron beam recrystallized sample with 350 keV protons.

Electron Channeling Pattern of SOI

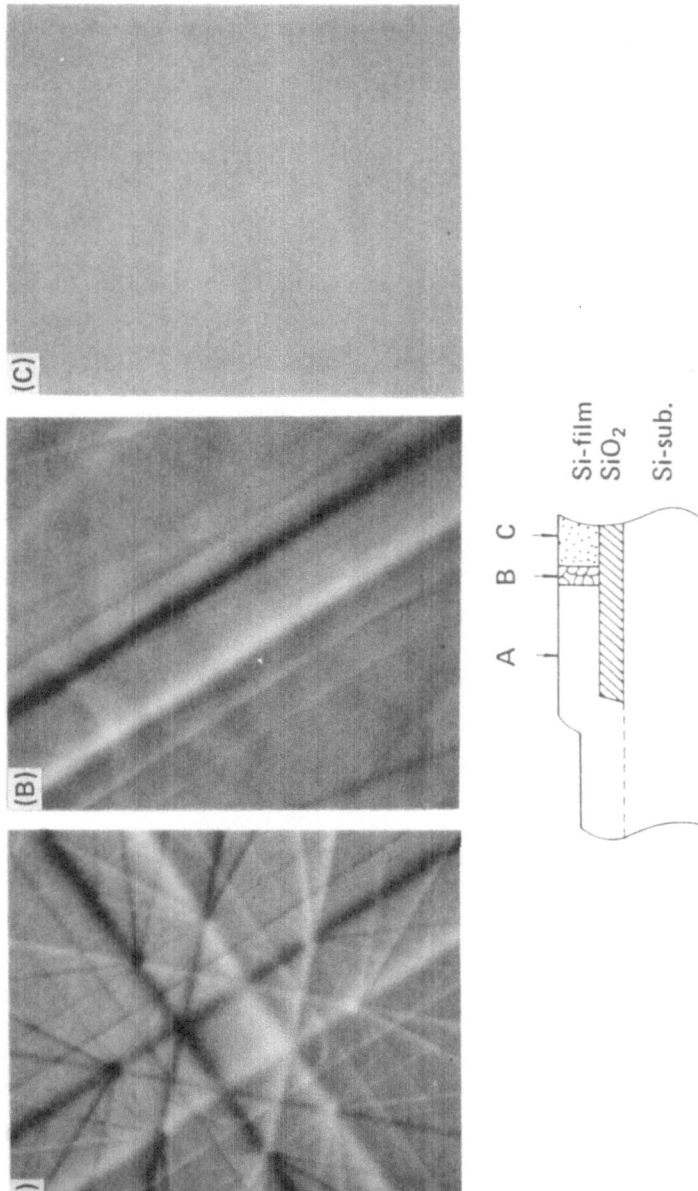

Fig. 9. Electron channeling patterns of the laterally seeded epitaxial growth sample; (A) complete epitaxial growth region, (B) transition region with (100) texture structure, and (C) without epitaxial growth region.

epitaxial growth sample. In this technique, the measuring area size was about 10 μm in diameter. In A region, complete epitaxial growth occurs from the seeded area, and channeling pattern shows (100) surface orientation. In B region, less than 40 μm from the SiO_2 pattern edge, texture structure appears, and only one of the (100) Bragg conditions is satisfied. On the contrary, in C region, no channeling patterns were observed. This is evidence that the epitaxial growth does not propagate to C region, and crystalline structure remains fine grained polycrystalline or amorphous.

5. Summary

Electron beam recrystallization of silicon films on silicon dioxide has been studied in three principal subjects. The first, grain boundaries were generated in the middle of the recrystallized region. It was confirmed from TEM observation that high density dislocations were generated at the strain concentrated region, and coalesced into the grain boundaries. The second, transient temperature profiles during the electron beam recrystallization was measured by a thermovision. The temperature profile was influenced by the beam scanning rate, and it has a convex tailing at the scanning rate of 5 mm/s. The third, laterally seeded epitaxial growth was investigated by using UHV deposited silicon films. Vertical seeding was accomplished during the silicon film deposition process, which expanded the margin of laterally seeded epitaxy with the electron beam, resulting in improvement of laterally epitaxial growth length and recrystallized film quality.

Acknowledgements
The authors would like to thank Dr. I. Higashinakagawa and T. Yoshii for their valuable discussions, and Y. Mikata for his technical support of UHV silicon deposition.

This work was performed under the management of the Research and Development Association for Future Electron Devices as a part of the Research and Development Project of Basic Technology for Future Industries sponsored by Agency of Industrial Science and Technology, MITI.

REFERENCES

1) K. Shibata, Y. Ohmura, T. Inoue, K. Kato, Y. Horiike, and M. Kashiwagi: Jpn. J. Appl. Phys. **22-1** (1983) 213.
2) T. Inoue, K. Shibata, K. Kato, Y. Mikata, and M. Kashiwagi: Extended Abstracts of the 15th Conference on Solid State Devices and Materials, Tokyo 1983, p. 93.
3) T. J. Stultz and J. F. Gibbons: *Laser and Electron-Beam Interactions with Solids* (North-Holland, New York, 1982) p. 499.
4) S. Kawamura, J. Sakurai, M. Nakano, and M. Takagi: Appl. Phys. Lett. **40** (1982) 394.
5) H. Ishiwara, M. Nakano, H. Yamamoto, and S. Furukawa: Jpn. J. Appl. Phys. **22-1** (1983) 607.

6) S. Banerjee and B. G. Streetman: J. Appl. Phys. **54** (1983) 2947.
7) L. Correra and G. G. Bentini: J. Appl. Phys. **54** (1983) 4330.
8) Y. Hayafuji, T. Yanada, S. Usui, S. Kawado, A. Shibata, N. Watanabe, M. Kikuchi, and K. E. Williams: Appl. Phys. Lett. **43** (1983) 473.

Silicon-on-Insulator: Its Technology and Applications, edited by S. Furukawa, pp. 41–46.
© KTK Scientific Publishers, Tokyo, 1985.

RECRYSTALLIZATION OF SOI STRUCTURES BY SPLIT LASER BEAM

N. AIZAKI

Fundamental Research Laboratories, NEC Corporation 1-1, Miyazaki Yonchome, Miyamae, Kawasaki 213, Japan

Abstract To obtain a large-area recrystallized single-crystal silicon layer on insulator, a birefringent quartz plate has been used to optimize the annealing laser beam shape. 20-μm-wide 1-mm-long single-crystal Si films on SiO_2 have been produced with a two-peak cw Ar laser beam. The single-scan recrystallized region width has been widened by multiple beam splitting. Using a four-peak beam, the resultant single-crystal region by multiple scan from the seed is a 100-μm-wide and 100-μm-long area with rare grain boundaries. This split beam method uses the stable and highly efficient TEM_{00} mode and needs no pre-patterned antireflection layers.

Recently, SOI(Silicon on Insulator) technology has been developed by many researchers.[1-9] SOI substrates with high-quality crystallinity have a potential to replace SOS substrates. Moreover, SOI technology is substantial to realize three-dimentional integrated circuits, which are highly effective to realize ultra-high density memory devices or multi-functional devices.

For application to three-dimentional device structure, the SOI fabrication method must not affect the characteristics of under-layer devices, which are already fabricated. Beam annealing method, laser or electron, is suitable for that purpose, because this kind of method can anneal only the surface of a specimen. The laser beam is considered to be more suitable than an electron beam, from the viewpoint that selective annealing is easily accomplished with patterned overlayers and no charge-up problem, which is inherent with an electron beam, exists.

However, the recrystallized single-crystal area is rather small for the LSI fabrication up until now. The essential discussion to obtain a large single-crystalline region is as follows. Grain growth direction must be controlled to start from the center of the scan width to either side, in order to suppress the competitive random nucleation from the outside polycrystallites. As a result of controlling grain growth, a single crystalline region is formed at the center region. The concave part of the liquid-solid interface line is necessary to realize that grain-growth direction pattern, because grain growth direction is normal to the liquid-solid interface line.

The liquid-solid interface line shape is determined by the laser beam shape and the laser power distribution. For instance, a crescent-shaped[5] or doughnut-shaped[7] laser beam is suitable to get the above-mentioned liquid-solid interface line shape, but the power efficiency is low when using those beam shapes and the doughnut beam shape is unstable. Another method to obtain that liquid-solid interface line shape is to use the stripe-patterned antireflection layer.[8] It requires a pre-patterned over-layer and aligned laser scanning.

This report presents an advanced split beam method to get a suitable and stable beam shape and the experimental results. It should be noted that this split beam method uses the stable fundamental TEM_{00} mode and needs no pre-patterned antireflection layers. Figure 1 is a diagramatic representation of this split beam method. 650-nm poly-Si was deposited by LPCVD on top of a 270-nm thick silicon oxide layer thermally formed on p-type (100) oriented Si. The silicon oxide layer was stripe-patterned

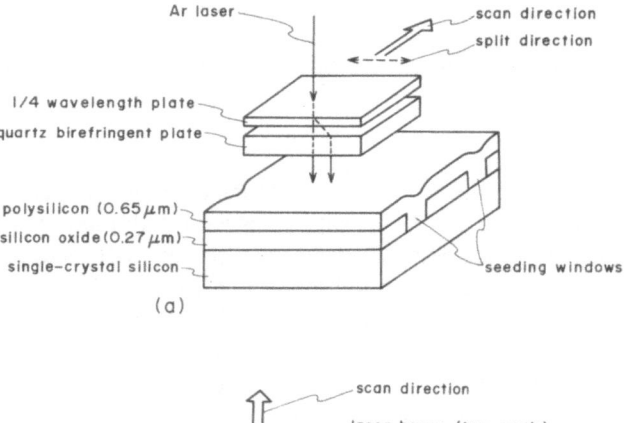

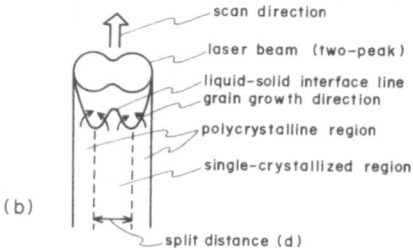

Fig. 1. Split beam method scheme: (a) splitting assembly and specimen structure, (b) diagramatic illustration of the laser beam profile and the liquid-solid interface line with the grain-growth direction.

to make seeding windows where poly-Si was in direct contact with substrate Si. Laser annealing scans were made across these seeding windows.

The cw Ar laser beam used in this experiment was split through a birefringent quartz plate and scanned in the normal direction to the split direction. Usually, the laser beam is linearly polarized at a resonator output. The laser beam was changed to a circular polarization using a quarter-wavelength plate, which is set in front of the splitting birefringent plate. Laser annealing was performed by using the split beam with a 2.5–5.0-W output power, a 10-mm/s scan speed, a 20–40-μm beam diameter and a 10–20-μm split distance. The sample temperature was kept at 300°C in air.

Figure 2 shows an SEM micrograph of the grain boundaries delineated by Secco etching, resulting from a split beam single scan with 4.6-W output power (P), a 10-mm/sec scan speed (v) and a 20-μm split distance (d). The SEM micrograph shows that the edge nucleation was suppressed and a 20-μm-wide 1-mm-long single-crystal region was obtained. Here the melt zone is supposed to have a concave-shaped region as shown in Fig.1(b) and a center crystallite of poly-Si plays a role of a seed, though there is no bulk silicon seed in the case of a lower power than the threshold power for seed region melting. The recrystallized single-crystal region width is equal to the beam split distance. This is understood by considering that the above-mentioned grain-growth direction started from the center region spread to the either side of concave region of the melt zone and the concave region width corresponds to the laser peak distance, which is equal to the laser beam split distance. Figure 3 shows the appropriate power

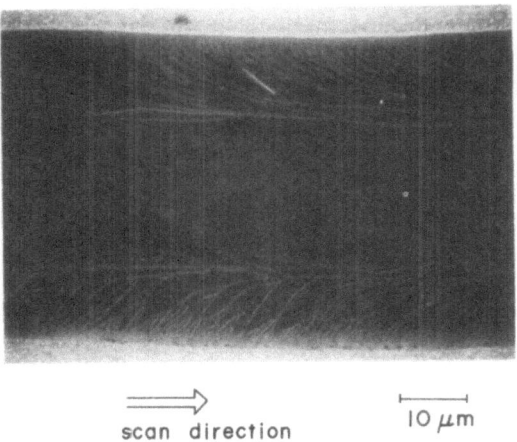

scan direction   10 μm

Fig. 2. SEM micrograph of the crystallized pattern obtained by a single laser scan following Secco etch. The center single-crystal region continues as long as 1 mm.

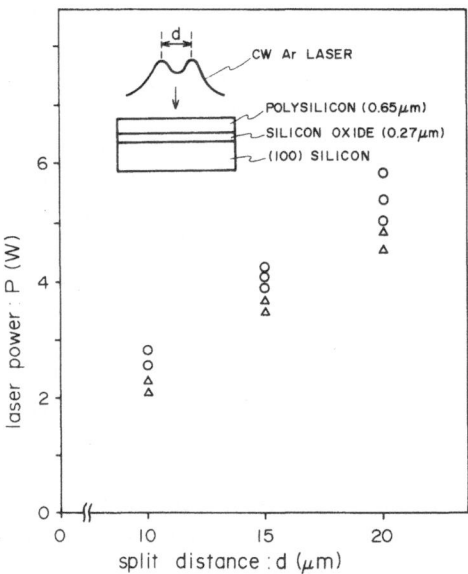

Fig. 3. Appropriate two-peak laser power region. (○:sufficient melting at the seed area. △:insufficient melting at the seed area.)

region as a function of the beam split distance.

Furthermore, this method has the potential to widen the single-scan recrystallized region width by multiple beam splitting. The multiple beam-splitting is accomplished as that the split beam obtained through the first birefringent quartz plate is passed through a quarter wave-length plate and then through the next birefringent quartz plate. The second birefringent plate has twice the thickness of the first birefringent plate, so that the resultant beam has four equally-separated peaks. The resultant recrystallized region width is three times as wide as the width of the recrystallized single-crystal region obtained with only the first birefringent plate. Figure 4 shows the Secco etched pattern obtained by a single scan with a four-peak laser beam. In this case, there are grain boundaries between individual bands of three single-crystal regions, which correspond to the concave parts of a liquid-solid interface line. The number of multiplication steps is limited by the laser power, because the resultant split beam must have sufficient power density to melt the poly-Si.

The above-mentioned results were all obtained with a lower laser power than the threshold power to melt the seed area completely. In the case of an appropriate laser power, the patterned seed area melts completely and the grain boundaries did not exist between individual single-crystal band regions obtained by a single-scan with a multi-peak beam and bet-

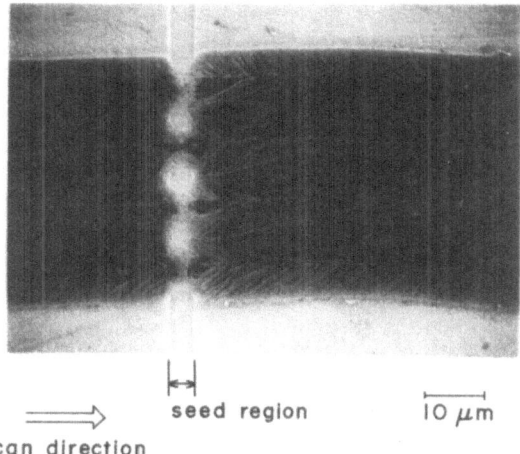

seed region   IO μm

scan direction

Fig. 4. SEM micrograph of the crystallization pattern obtained in a single scan with a four-peak laser beam following Secco etching. The seeding window is seen at the left. Annealing conditions are $P = 4.6$ W, $v = 10$ mm/s and $d = 10$ μm.

ween individual scan regions obtained by multiple scans. The obtained large single-crystal layer had the same crystallographical direction as the seed substrate. The resultant large single-crystalline area is shown in Fig. 5, where a 100-μm-wide and 100-μm-long area is recrystallized into single crystal with rare grain boundaries using a four-peak split beam. In that

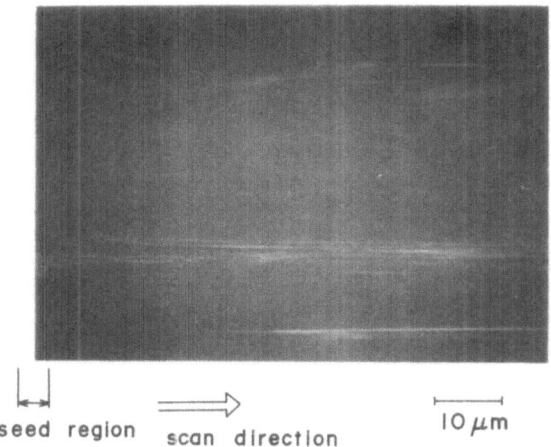

seed region   scan direction   IO μm

Fig. 5. SEM micrograph of the large-area recrystallized single-crystal region obtained by multiple scans with a four-peak beam following Secco etch. The seeding window is seen at the left. Annealing conditions are $P = 4.8$ W, $v = 10$ mm/s, $d = 10$ μm and v (scan pitch) $= 30$ μm.

area, the crystallographical direction was checked by the ECP (electron channeling pattern) method to be the same as the seed substrate direction at any point.

In conclusion, using a novel split-beam method with the birefringent quartz plate, the grain size of laser-recrystallized poly-Si on SiO_2 was increased with stability and without any complexity in scanning or substrate structure.

Acknowledgements

The author whould like to thank Drs. N. Kawamura, T. Kawamura and H. Tsuya for their interest and support in this research. He also wishes to thank his collegues for their useful discussions and equipment reconstruction.

This work was performed under the management of R & D Association for Future Electron Devices as a part of the R & D Project of Basic Technology for Future Industries sponsored by the Agency of Industrial Science and Technology, MITI.

REFERENCES

1) M. W. Geis, D. C. Flanders, and H. I. Smith: Appl. Phys. Lett. **35**(1979) 71.
2) J. F. Gibbons, K. F. Lee, T. J. Magee, J. Peng, and R. Ormond: Appl. Phys. Lett. **34**(1979) 831.
3) M. Tamura, H. Tamura, and T. Tokuyama: Jpn. J. Appl. Phys. **19**(1980) L23.
4) D. K. Biegelsen, N. M. Johnson, D. J. Bartelink, and M. D. Moyer: Appl. Phys. Lett. **38**(1981) 150.
5) T. J. Stultz and J. F. Gibbons: Appl. Phys. Lett. **39**(1981) 498.
6) B. -Y. Tsaur, J. C. C. Fan, M. W. Geis, D. J. Silversmith, and R. W. Mountain: Appl. Phys. Lett. **39**(1981) 561.
7) S. Kawamura, J. Sakurai, M. Nakano, and M. Takagi: Appl. Phys. Lett. **40**(1982) 394.
8) J. P. Colinge, E. Demoulin, D. Bensahel, and G. Auvert: *Proc. 14th Conf. Solid State Devices, Tokyo, 1982*, Jpn. J. Appl. Phys. **22**(1983) Suppl. 22-1.
9) H. Ishiwara, M. Nakano, H. Yamamoto, and S. Furukawa, *Proc. 14th Conf. Solid State Devices, Tokyo, 1982*, Jpn. J. Appl. Phys. **22**(1983) Suppl. 22-1.

Silicon-on-Insulator: Its Technology and Applications, edited by S. Furukawa, pp. 47–66.
© KTK Scientific Publishers, Tokyo, 1985.

NUCLEATION AND CRYSTAL GROWTH CHARACTERISTICS IN ENERGY BEAM CRYSTALLIZATION OF SILICON ISLANDS

K. KUGIMIYA, S. AKIYAMA, and N. YOSHII

Central Res. Lab., Matsushita Elec. Ind. Co., Ltd., Moriguchi, Osaka 570, Japan

Abstract Nucleation and crystal growth controls in recrystallization of isolated Si islands embedded in insulators have been studied.

It was observed that isolated single-scannings of polysilicon films by energy beams resulted in the preferred orientation of (110) surface and $\langle 111 \rangle \sim \langle 113 \rangle$ growth directions. Overlapped multiscannings were destructive to this preferred orientation and resulted in random orientation which was due to unstable and random nucleation at Si/melt-Si/SiO₂ interfaces and also due to crystal break-ups caused by large stresses.

In crystallization of islands, the (110) and $\langle 111 \rangle$ preferred orientation was also observed. The (110) surface orientation was assumed to reflect the high (110) anisotropy of as-deposited LPCVD polysilicon films and/or the stress enhanced growth accompanied with (112) pseudo-twins. A model was constructed and showed that pseudo-twins extending to $\langle 111 \rangle$ directions were nucleation sites for the very fast growth and accommodated impurities like oxygen, effectively reducing stresses. At melt-Si/SiO₂ interfaces, nucleation was almost suppressed reflecting preferred temperature profiles.

These observations pointed out that islands embedded in insulators with grooves and traced by isolated single scannings resulted in the best grain boundary control and better single crystal islands.

1. Introduction

Recently various efforts have been focused on fabricating SOI devices and multilayer (3 dimensional) devices. It has been reported that ring oscillators, dividers and a driver of a liquid crystal display were successively fabricated in SOI substrates and showed fair device performances.[1-6] Simple devices of stacked CMOS[7] and J-CMOS[8] were also fabricated and showed future possibilities of 3D devices. However, all of these devices were fabricated in polysilicon films or islands with relatively large grain sizes, up to some 10 μm. Distributions of grain boundaries, grain sizes and crystal orientations were mostly uncontrolled,[5] and were subject to difficulties of controlling electronic properties, i.e. poor reproducibility. Large ΔV_T's were resulted from unstable N_{SS} and gate oxide thickness which were caused by different crystal orientations probably enhanced by surface damages and irregularities. Large Δg_m's were associated with the enhanced impurity diffusion and mobility instabilities at grain boundaries or strained areas. With all of these difficulties we can still expect fair and even device properties from some experimental evidences and statistical expectation when devices are based on large design rules, say to 4 μm rule, and are fabricated in polysilicon substrates with large grain sizes barely exceeding 0.5 μm. This is simply because electronic properties of polysilicon films become quite close to those of bulk silicons when grain sizes are larger and because devices contain sufficient number of grains to equallize uneven properties derived from grains. We can also expect good device properties when devices are built in large monograins, say 100 μm.[9] These approaches are clearly limited and can not be applied to conventional IC production.

In order to introduce the SOI technology successively into the existing or future semiconductor processes, the technology must be superior to the competitive technologies such as submicron and hybridized VLSI's, i.e. SOI or 3D devices must furnish VLSI level high densities and uniformities over entire wafer areas. To this end we have to develop single-crystal layers on insulators with (1) complete crystal orientation control, (2) complete grain boundary control, and (3) high crystal quality without defects, over entire wafer surfaces without sacrificing active device areas. To realize these requirements, many improvements have been reported since the application of laser annealing technology for the recrystallization of polysilicon films and islands. For the stable and continued crystal growth, various beam shaping methods[10-13] have been tried out and further for reducing stresses, various island formations and shapes[14-19] have been studied. Crystal orientation controls by seeding[20,21] and graphoepitaxy[22] have been studied. Yet none is successful to offer substrates qualified for SOI or 3D devices in the VLSI level at present.

This paper describes the basic experimental observations on nucleation and crystal growth of polysilicons melted by laser and electron beam irradiations and discusses about the possibility of controlling crystal orientations, grain boundaries and crystal defects.

2. Experimental Method

2.1 Energy deposition sources

C.W. scanning laser and electron beams were used for irradiating polysilicon specimens.

The laser beam (LB) irradiation apparatus[23] consists of a CW Ar laser (Coherent CR-18), a shutter, a $5 \times$ expander, defocus lenses, apertures, final focusing lenses ($50 \sim 250$ mm^f), a monitor TV and X-Y scanning stages with a 3″ wafer holder. The X-Y stages driven by a desktop computer (HP 35A) have a positioning accuracy of 2 μm. Specimens are heated up to 400°C and are monitored directly through final focussing lenses.

Beam shapes were measured[24] by a photo diode with an 1 μm$^\phi$ pinhole cover at a specimen position on the X-Y stages after reducing the primary laser beam to 1/100 by a beam splitter. Diode outputs were plotted out at every 2 μm increment of the X-Y stages with colour coding corresponding to the output intensities. A gaussian profile was observed as shown in Fig.1(a). A letter p in the figure indicates a peak position of a beam. A beam diameter at 1/e was about 20 μm$^\phi$ and corresponded to a melt width of a 0.5 μm polysilicon film on 1 μm SiO$_2$ (Fig.2.1) at an optimum irradiation condition. Irradiation conditions were; input power 3–5W, scanning speed 100 mm/sec and specimen temperature 320°C.

The electron beam (EB) irradiation apparatus is the model JEBA-011G (JEOL) which is capable to deliver powers up to 30 mA/10 keV and to scan up to 5000 mm/sec at pitches of minimum of 1 μm to 99 μm. The apparatus is furnished with defocussing lenses to change beam diameters and with stigma lenses to obtain line beams. Through a load-locked sub-chamber, 3″ specimen wafers are set in position on a specimen holder

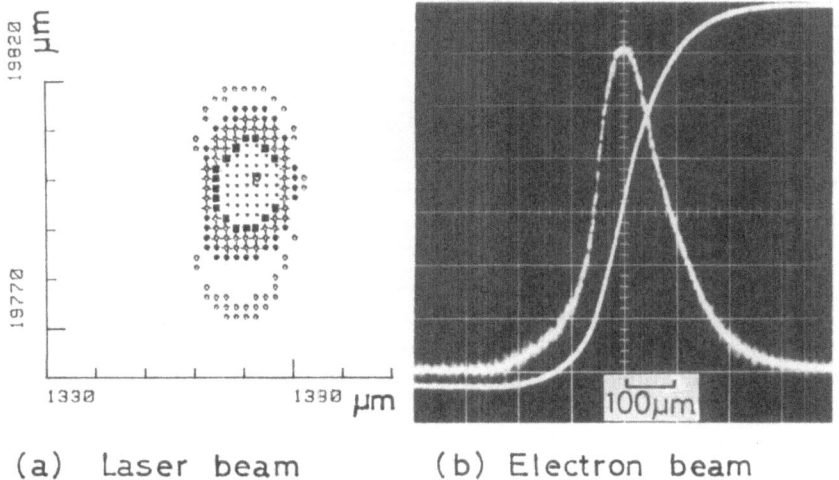

(a)   Laser beam   (b) Electron beam

Fig. 1. Beam profiles of laser and electron beams.

K. Kugiyama *et al.*

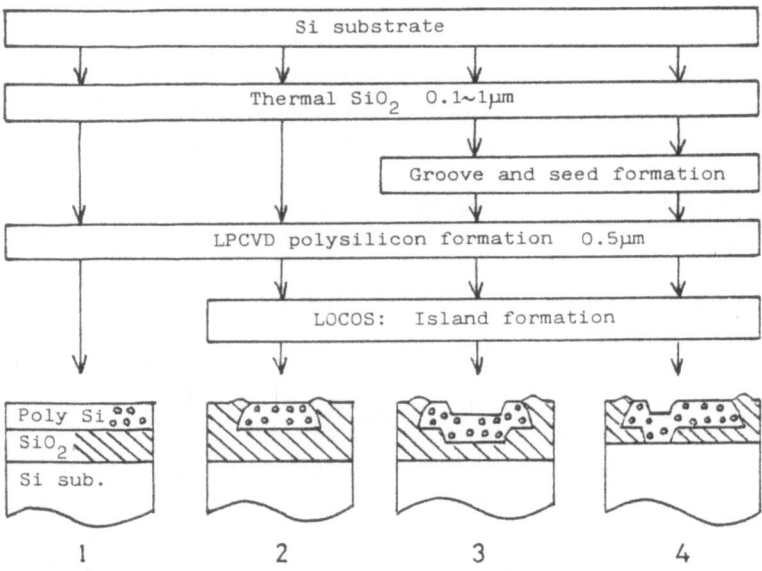

Fig. 2. Sample preparation.

which can be heated up to 800°C by a doubly wound Ta wire heater in a vacuum of 2×10^{-6} Torr. Specimen currents are always monitored *in situ* as voltages across an 1 kΩ resistor and usually 60-70% of total emission currents. For beam shape monitoring, a miniature Faraday cup (1 mm$^\phi$) is furnished. In Fig.1(b) a total beam current passed through an aperture of a Faraday cup and its derivative, a Gaussian beam profile, are shown. A beam diameter at $1/e$ was about 200 μm whereas a melt width was only 80 μm at an optimum irradiation condition; input power 2.3 mA/10 keV, scanning speed 500 mm/sec and specimen temperature 330°C on a polysilicon film specimen (Fig.2.1). A smaller melt-width/beam-diameter ratio of an EB trace compared with that of an LB trace was mainly due to a broader beam shape (large varian) and a lower peak height of an electron beam.

2.2 Sample preparation and analyses method

All specimens used in this experiment are tabulated in Fig.2. Thermal oxide layers of 1 μm thick were grown on 3″ (100) or (111) Si wafers and polysilicon films of 0.5 μm thick were further deposited by LPCVD. The LPCVD polysilicon layers were grown by thermal decomposition of SiH₄ gas with 80% He dilution at 0.8 Torr and 610°C. The grown layers consisted of small grains of about 500 Å and exhibited very high, nearly 100% (110) orientation as shown in Fig.6. This completed specimen 1, continuous polysilicon films covering entire wafer surfaces. For the preparation of specimen 2, LPCVD Si_3N_4 layers were further deposited on specimen 1 and selectively etched away by a conventional photolithographic technology leaving Si_3N_4 layers corresponding to islands. Polysilicon films were then etched away by about a half of their thickness so that top surfaces became flat after oxidation. Si_3N_4 layers were then removed. For the preparation of specimens 3 and 4, grooves and seed vias were made by wet etching before the LPCVD polysilicon deposition process.

Beam (LB and EB) irradiations were generally done at optimum conditions as described before. Overlappings were typically 40–70% except single scanning.

Recrystallized samples were first examined by a bright field (BF) and dark field (DF) optical microscope (OM). Samples were then etched to delineate grain boundaries and defects for further examination by an OM. Since DF images corresponded quite well with BF images of etched surfaces, DF observations were most frequently applied to see the microstructure. For transmission electron microscopic (TEM) observations, recrystallized samples were thinned by etching. Etching from the backside of the samples was automatically stopped at Si/SiO_2 interfaces. Microstructures of crystals were then analysed through thick (1 μm) SiO_2 layers by JEM200CX (JEOL) at 200kV acceleration.

Anisotropy measurements were made through X-ray powder patterns. Samples were rotated at more than 1000 rpm to increase inspection areas for better statistics and reproducibility. It has been experimentally known that without rotation reproducibility became poorer if grain sizes became larger than about 10 μm[25].

3. Experimental Results and Discussion

3.1 Recrystallization of continuous polysilicon films
3.1.1 Single isolated scanning

Figure 3 shows a typical trace of a polysilicon film (Fig. 2-1) by LB single isolated scanning. It showed a typical chevron pattern consisted of mainly three areas, central area which contained large elongated crystals with their long axes aligned to a laser scan direction, intermediate area which contained medium size crystals with their elongated growth directions inclined to a laser scan direction, and peripheral area which contained mainly small grains with their growth directions almost perpendicular to a laser scan direction. TEM analyses showed that the most frequently observed crystal growth directions and surfaces were ⟨111⟩, ⟨112⟩ and ⟨113⟩ directions and (110) surfaces in the central area.[26] Closer examination showed that most of these large elongated crystals were initiated at the scan boundary of the peripheral area. In the intermediate and peripheral

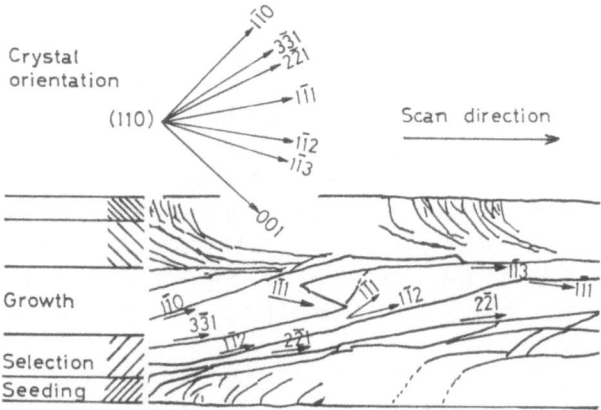

Fig. 3.   A single isolated scanning of a polysilicon film.

areas, small crystallites also showed frequent (110) free surfaces but their
growth directions were nearly random. There were quite a few crystallites
which were initiated off from scan boundaries. These crystallites were
assumingly made by the random nucleation. These observations clearly
indicated that

(1)   the nucleation took place on polysilicons at scan boundaries which
had high (110) anisotropy as shown in Fig.6, thus crystallites started to
grow with (110) surfaces in the peripheral area or seeding area;

(2)   crystallites with the preferred growth directions, $\langle 111 \rangle \sim \langle 113 \rangle$
were kept growing along the thermal gradient and undesirable ones were
weeded out in the intermediate area or selection area;

(3)   finally selected crystals expanded and grew along the preferred
directions, $\langle 111 \rangle \sim \langle 113 \rangle$, with their axes in parallel with a scan direction
in the central area or growth area.
Since four $\langle 111 \rangle \sim \langle 113 \rangle$ directions are present in (110) planes, crystals
with the (110) preferred orientation could grow along the chevron pattern
with their growth axes matching with some of the $\langle 111 \rangle \sim \langle 113 \rangle$ directions. More discussion about the preferred orientation is made in the follow-
ing section.

Figure 4 shows a BF image of a recrystallized and etched film after
scanning of a line EB with an 1 mm wide trapezoidal beam profile.[27]
Thin crystallites grew and elongated along a scan direction up to a few
mm, and many branched out frequently into several crystallites running
almost parallel with each other. The branched crystallites were separated
by subgrain boundaries as indicated by a circle in the figure. Crystallites
clearly showed periodical ridges at boundaries indicating unstable growth
nature.

TEM analyses of about some ten neighbouring grains showed that

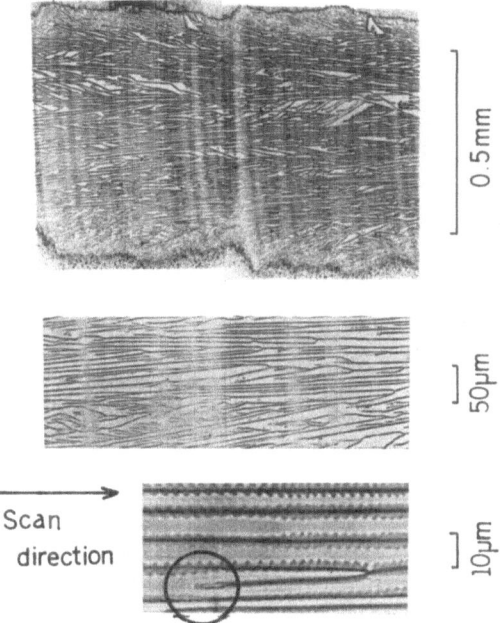

Scan
direction

Fig. 4. 1 mm wide EB line beam scanning with a trapezoidal intensity profile.

crystallites had nearly (110) preferred orientation and grew almost in parallel
with ⟨111⟩ direction. The observed (110) planes were within ±5° of
specimen surfaces and many were with the ⟨311⟩ zone axis which lay
down about 80° from a scan direction in the (110) surface plane. No
appreciable defect was observed except grain boundaries and subboundaries.

The crystal break-ups observed in parallel with a scan direction indicated that a strong strain field was built up during freezing and it had
a mirror symmetry along a scan direction. An observed stress calculated
from a warpage of a specimen was far greater than 10^{10} dyne/cm^2 which
was even greater than the stress associated with the multiscanning.[28]

3.1.2 *Overlapped multiscanning*

Effects of overlapping were shown in Figs.5 and 6. Straight growth
in a scan direction was observed when overlapping was small generally
about 50% or less (Fig.5 c, d). Vertical growth almost perpendicular to
a scan direction was clearly shown when overlapping was larger than 70%
(Fig.5 a, b). Chevron-like growth by LB scans was simply made by scanning both ways. Overlapped-multi scanning in a single direction resulted
in a simple vertical growth (Fig.5e by EB scans). This vertical growth
was due to the fact that seedings of succeeding scans took place at the
seeding or selection areas of outerside where temperature gradients or crystal

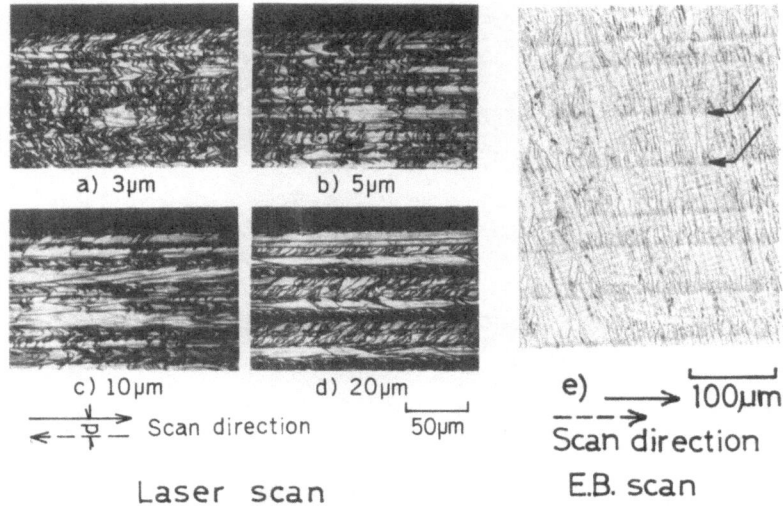

Fig. 5. Crystal growth habits in overlapped multiscanning.
Beam diameter: Laser 20 μm; Electron 100 μm.

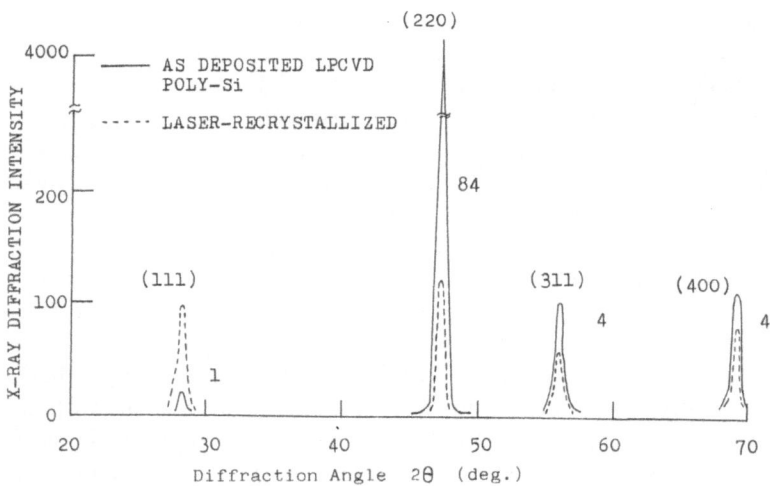

Fig. 6. X-ray diffraction patterns normallized by the Si powder pattern.

growths were nearly perpendicular to scan directions (those areas of succeeding scan sides were completely melted). Closer examination of the boundaries of remelt-Si pointed by arrows, revealed the presence cf frequent crystal break-ups and random nucleation which were indications of large strains at remelt areas.

Figure 6 shows X-ray diffraction patterns of an as-deposited polysilicon film and an overlapped (70%) multiscanned specimen. Although a polysilicon film showed very high degree, almost 100%, of the (110) orientation and although a single scanned specimen had the preferred (110) orientation in the growth area, an overlapped multiscanned specimen completely lost the (110) orientation. This was due to the fact that seeding took place at the seeding and selection areas where the selection was nct completed and that crystal break-ups and random nucleation frequently took place at remelt-Si interfaces as described above.

3.2 Recrystallization of polysilicon islands
3.2.1 Recrystallization of simple islands

Figure 7 shows typical recrystallization characteristics of polysilicon islands (Fig.2.2) by EB scanning. LB scanning also yielded quite the same results. When an irradiating power was low, quite obvious regrowth took place only at the edges and slight regrowth in the center of islands, indicating lower temperatures in the center. When a power was optimum, large crystals were grown and filled islands. Notice that grain boundaries met at one point almost in the center of each square island as pointed out by an arrow. The same points were also present at the heads of long

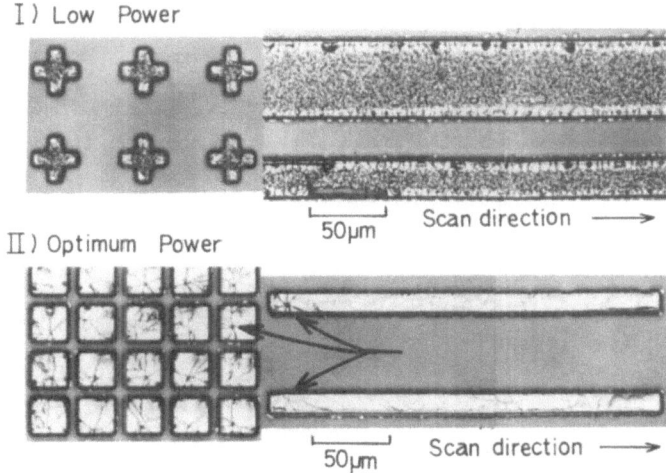

Fig. 7. Nucleation and crystal growth of LOCOS poly-silicon islands by EB scanning.

islands and large single crystals initiated at the points extending to the rest of islands. These points were clearly nucleation points since temperatures were the lowest in the center thus initiating growth sites and since several grain boundaries did not likely meet at one point in most of the islands unless grain growth started there. Also there was almost no nucleation at the SiO_2 edges. Simulated temperature profiles of an island in Fig.8[29] also showed that the center of an island had the lowest temperature and was the nucleation point. It was clear that, in general, Si substrates or polysilicon substrates below islands acted as heat sinks and that substrate areas at peripheries of islands were heated up and acted as virtual heat barriers.

Similar temperature profiles were expected for EB irradiation. Although energy depositions by laser and electron beams were different, exponential and Gaussian distributions respectively, temperature profiles resulted in islands became quite the same as schematically illustrated in Fig.9, since the heated SiO_2 surrounding islands acted as the virtual heat barriers in the EB case instead of the heated substrate areas in the LB case. This resulted in the same recrystallization characteristics by either LB or EB irradiations as shown in Fig.7.

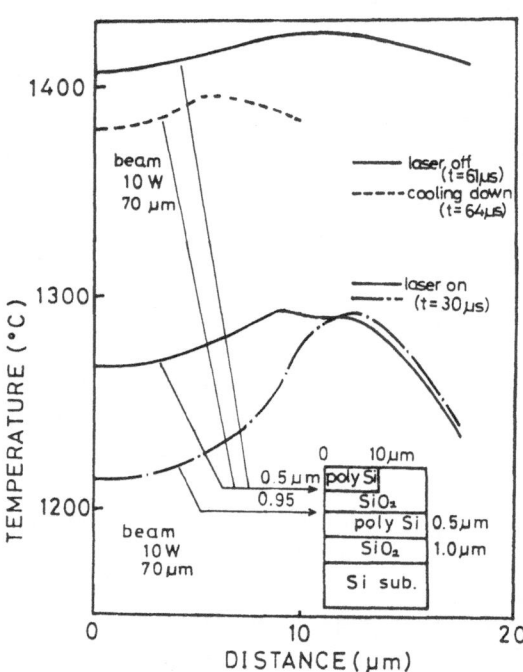

Fig. 8.   Simulated temperature profiles of an island embedded in an insulator. Beam: CW Argon laser.

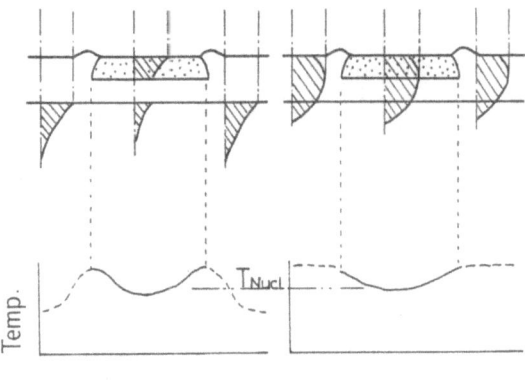

a) Laser beam   b) Electron beam

Fig. 9.   Energy deppositions and temperature profiles at island interfaces.

A TEM photograph of Fig.10 shows a closer view of a nucleation point.[27] White broad lines were unexpectedly etched cracks and grain boundaries through a broken oxide layer caused by an excess residual stress during a specimen preparation. TEM analyses showed that all crystallites in the island had nearly (110) surfaces except one with (113) surface at an area where the crystallite was frozen at the last moment of the recrystallization process. The thickness of the area was an indication since the last frozen areas of islands were generally much thicker than other areas and sometimes made protrusions like hillocks. Thus slower cooling or surface irregularities might complicate the crystal growth. Closer examination of Fig.10b revealed the presence of small dislocation networks a, pseudotwins b in $\langle 111 \rangle$ direction and c in $\langle 110 \rangle$ direction, and several grain boundaries (dashed lines) extending outward from a nucleation point. Preferred growth directions could not be determined in the regrowth of the square island since $\langle 111 \rangle \sim \langle 113 \rangle$ directions appeared with a high symmetry in (110) plane.

The same results were also observed in recrystallized long islands (Fig.11). Major crystal surfaces were again (110) planes and the ones with the preferred $\langle 111 \rangle$ growth direction expanded to cover the rests of the islands as they grew up. This (110) and $\langle 111 \rangle$ preferred crystal orientation was the same as in polysilicon films described above. Two kinds of pseudotwins were observed in the preferred (110) orientation. Major ones ran along $\langle 111 \rangle$ directions with (112) boundaries and minor along $\langle 110 \rangle$ with either (112) or (111) boundaries as shown in Figs. 10 and 11. Another defect, may be slipline, running along $\langle 001 \rangle$ and about 5° off from a scan direction (Fig.11) was also observed. However slips and stacking faults

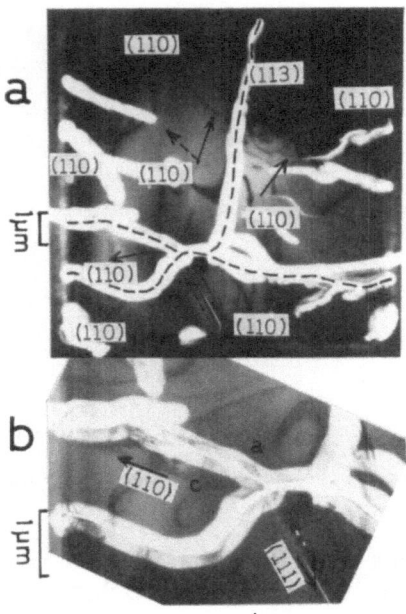

Fig. 10.   A nucleation point of a recrystallized island.

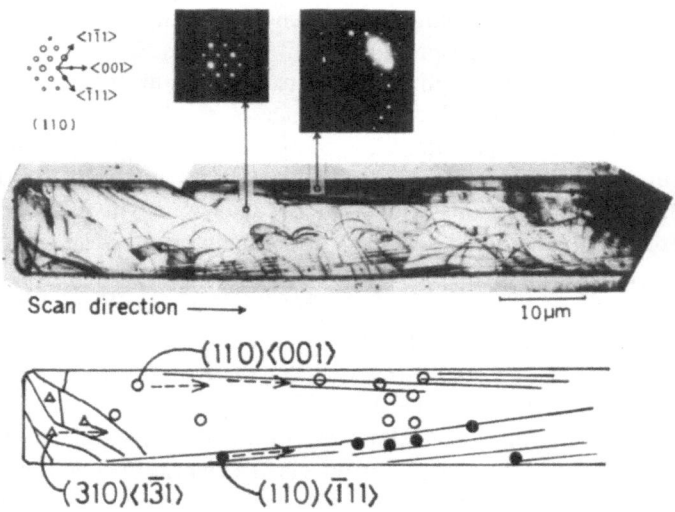

Fig. 11.   Recrystallization characteristics of an island embedded in an insulator.

normally in (111) planes were scarcely observed.

3.2.2 Recrystallization of islands with grooves

DF images in Fig.12 shows typical growth characteristics of islands with grooves (Fig.2.3) made by LB and EB scannings. All specimens showed large continuous single crystal formation in grooves and microcrystal growth on terrace areas which collared around grooves. An island even with an irregular shape, Fig.12a, showed complete single crystal formation filling only grooves with less number of crystal defects. LB scanning was made in parallel with grooves. When a single isolated LB scanning was made in perpendicular to grooves, Fig.12b, mainly two large crystals were grown forming grain boundaries in the center of a scan trace and smaller crystals remained on terrace areas as schematically shown in an inset of Fig.12b, Occasionally large crystals grew up extending over terraces but no crystal grew down into the grooves. Wider recrystallized areas on terraces indicated that temperatures were higher on terrace areas than in grooves as expected. Heat flows were schematically illustrated in Fig.13 assuming the substrate temperature profile of Fig.8. Maximum heat flow during an initial stage of a cooling period was expected at an edge of a groove, since a substrate temperature was the lowest in the center of an island and since heat flow was larger in a groove due to a thinner oxide layer below a groove with extra heat flows to the side oxide layers.

It was thus quite clear that two large crystals nucleated on the corners 2, 3 in Fig.12b and grew inward along the groove edges 2~3 where temperatures were lowest, and that numerous nucleations took place at the terrace edges 4~5 due to the fast heat dissipation to the frozen silicon in grooves. Occasional overgrowths onto terrace areas from grooves were probably due to that the preferable growth directions were also present when crystal growths bent over the edges 4~5. It was also noted that at the edges 4~5, numerous small crystals and the large two crystals formed clear boundaries exactly matching with groove edges. This matching indicated that the preferable crystal growth directions were mostly unidirectional and that crystals grew straight hitting the edges and were stopped with slight growth, up slightly over the slopes of the edges.

When a large island or a polysilicon film with grooves (Fig.12c) were recrystallized by EB scanning, large crystal growth was also consistently observed. The continuous growth was broken by occasional boundary formation at every 30~300 μm. By LB scanning the same results were also obtained. For a control of grain boundaries, islands were thus preferably shorter than 30 μm at least in this experimental condition.

A preliminary TEM examination showed the frequent occurrence of (110) surfaces. More studies are necessary to determine the preferred crystal orientation and the growth direction.

3.2.3 Recrystallization of seeded polysilicon islands

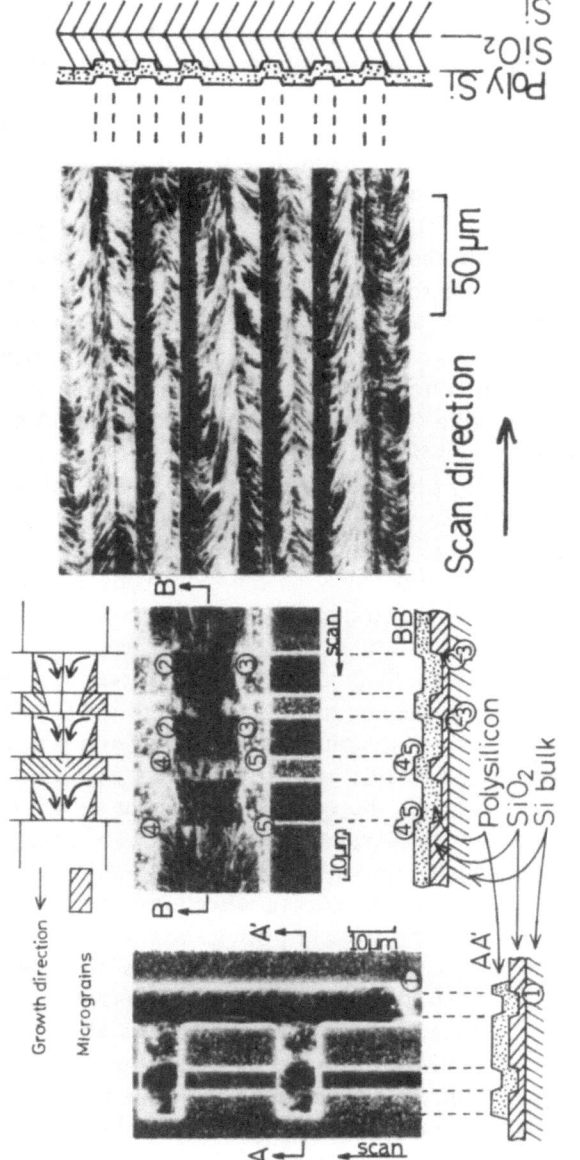

Fig. 12.   Nucleation and crystal growth of polysilicon islands with grooves.
a, c) Parallel scanning. b) Perpendicular scanning.

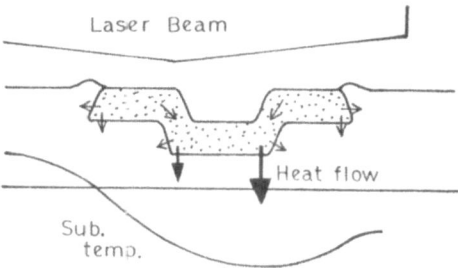

Fig. 13.   Heat flows and a temperature profile of an island with terraces.

An orientation control of islands was realized by simply seeding to substrates through vias (Fig.2.4). A large single crystal (Fig.14) grew up to 250 μm long with very few defects inheriting a seed orientation of (111) surface and $\langle 110 \rangle$ growth direction as shown by an electron channeling pattern in Fig.14b. When the growth was terminated, (110) surfaces appeared frequently. Crystal growth over SiO_2 layers from vias was not only in the scan direction but also to the side, expanding by 30 ~ 40° to edges of SiO_2. No nucleation was also observed at SiO_2 edges.

Seeding was also sucessive on (100) and (110) surfaces. In seeded recrystallization, however, precise power controls were necessary for the successive growth. Optimum melting of both seeding and growth areas on SiO_2 layers only resulted with a very narrow power window. Thickness of oxide layer was limited to about 1000 Å at most for the stable con-

(a) Laser Recrystallized Si Layer of Seeded Island   10μm

$\theta = 10°$

(b) Electron Channeling Pattern   (111)

Fig. 14.   Recrystallization of a seeded polysilicon island.

tinued growth. Cover layers might be necessary to homogenize heat distribution and to reduce local overheating, as heat sinks or selective power reflectors.

3.3 Some discussion on crystal orientations and defects

Recrystallization of polysilicon films by isolated single-scannings or of islands by either LB or EB scannings consistently showed the (110) and ⟨111⟩ preferred growth orientation and presumably (112) pseudo-twins extending in ⟨111⟩ directions.

The preferred orientation of (110) has been reported by Nishioka *et. al.*[30] on laser recrystallization of Ge films though it was also accompanied by the occasional (111) appearance. Gale[31] *et al.* have also reported the (110) and ⟨001⟩ preferred growth. For recrystallization of silicon, Geis *et al.*[32, 33] have reported the (100) preferred orientation when encapsulant layers were present, and no preferred orientation when no encapsulant layer and no oxygen in an atmosphere was present. The weak (110) preferred orientation with (100) presence was also reported by Nishimura.[34] All of these reported orientations were inconsistent and the difference from our results was also ambiguous. This was probably due to the different experimental conditions such as the orientation of as-deposited polysilicon, cooling speeds and residual stresses at the interface with insulators.

The (110) preferred orientation observed in our experiments was probably either due to (1) the high (110) anisotropy of as-deposited LPCVD polysilicon layers or to (2) some Si/SiO_2 interfacial anisotropy associated with the large thermal and solidification stresses coupled with oxygen in an atmosphere, or both.

In the first case, it was assumed that some embryos or nuclei still remained unmelted at the bottom interfaces with oxide layers at the optimum recrystallization condition. At slightly higher power conditions exceeding the optimum, embryos or nuclei might have disappeared resulting in the random orientation. However, at the excess power conditions, melts were blown off and crystallization was very poor as well known. Thus the hypothesis could not be proved.

In the second case, (110) and ⟨111⟩~⟨112⟩ preferred growth was assumingly stabilized as reported in the ribbon growth of Si[35] with a complex interaction among a fast growth rate, large stresses and an inclusion of impurities like oxygen. The large stresses associated in the growth were obviously one of the most important factors influencing growth habits by the fast scannings, which habits were quite the opposite from the epitaxial regrowth of amorphous Si surfaces showing that (100) was the fastest regrowth plane and (111) the slowest.[36]

From the frequent occurrence of pseudotwins in (112) plane in the direction of ⟨111⟩ associated with the (110) preferred orientation, a growth model was introduced as schematically explained in Fig. 15. Provided that the (110) preferred planes were free surfaces in the growth, main growth directions would be either ⟨100⟩, ⟨110⟩, ⟨111⟩ or ⟨112⟩ directions. Since twins or stacking faults were expected to form in (111) planes for reduction of stresses (and nucleation sites for easier growth), and since they would grow in (111) planes once they formed, growths in ⟨100⟩ and ⟨110⟩ directions which are inclined to ⟨111⟩ directions or (111) planes were thus ruled out.

When the growth was assumed to be in ⟨$\bar{1}\bar{1}2$⟩ direction, twins were expected to grow favourably nucleating on twin boundaries as shown in the figure. However initiation of twins at the point a should be accompanied by defect formation and impurity segregations. After the initiation, quite stable and normal twin boundaries were formed if twin thicknesses consisted of multiple of three dense (111) layers, and if not, pseudotwin boundaries were formed probably including impurities.

When the growth was assumed to be in ⟨$\bar{1}\bar{1}\bar{1}$⟩ direction, twins were also expected to form naturally at point b as shown in Fig. 15 and twin formation energy was obviously much less than that at point a. As twins met on ($\bar{1}\bar{1}\bar{1}$) planes, they formed pseudotwin boundaries with impurity segregation like oxygen. The pseudotwin boundaries served as easier growth sites and stress relaxation sites.[37] The boundaries thus expected to grow along ⟨$\bar{1}\bar{1}\bar{1}$⟩ direction. The boundary surfaces were likely in ($\bar{1}\bar{1}2$) planes since an atomic arrangement with oxygen intercorporation in the planes was more favourable than in other main mirror planes (110) and since the boundary areas were minimized due to the orthogonal relation of ($\bar{1}\bar{1}2$) and ($1\bar{1}0$). Quite a few boundaries were observed, sometimes several for one micron width, and about $2 \times 10^{18}/cm^3$ of oxygen was measured by SIMS, which was close to saturation. Thus many growth sites were offered by the pseudotwin boundaries for the fast growth and oxygen con-

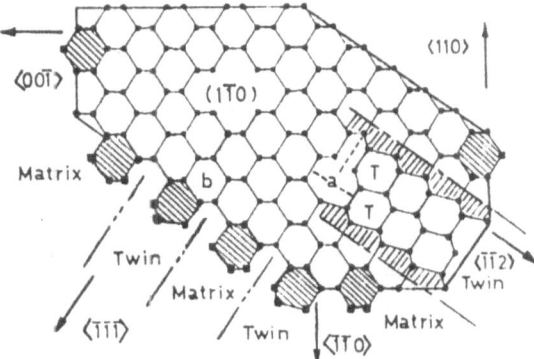

Fig. 15. Pseudotwin growth in ($1\bar{1}0$) surface.

centration was high enough for the enhanced segregation and the fault growth.[38] The stabilization of the faults by carbon and oxygen segregation was also suggested by Tan *et al.*[39]

Thus growth in ⟨111⟩ directions was the most favoured in the very fast growth by LB or EB scannings due to the presence of the easier growth sites. Growth in ⟨112⟩ direction was also favourable but more favoured in the slower growth like the ribbon crystal pulling.[35] The model was schematic but explained the main features of the experimental observations.

4. Summary

(110) and ⟨111⟩ ~ ⟨113⟩ preferred orientations in crystal growth were observed. The preferred orientations, though not so perfect, were consistently observed either on recrystallized films traced by isolated LB and EB single-scannings or on recrystallized islands embedded in insulators. The preferred orientations were assumed due to the inherent nature of (110) anisotropy of as-deposited polysilicon films or due to the stress relaxation incorporated with impurities like oxygen. A proposed growth model could explain the ⟨111⟩ and ⟨112⟩ growth preference and the formation of pseudotwins in (112) planes extending into ⟨111⟩ directions which served as the first growth sites and also as stress relaxation sites.

Overlapped multi-scannings, however, destroyed the preferred orientation probably due to the randum nucleation and crystal break-ups at remelt-Si/Si interfaces, both induced by the large stresses associated with recrystallization.

At melt-Si/SiO$_2$ interfaces, however, nucleation was almost suppressed. Thus complete single crystal islands were obtained when Si islands in grooves of SiO$_2$ were entirely covered by energy beams which diameters were wider than island widths and when melted zones completely filling grooves scanned along islands from one end to the other. Analyses showed a preferable temperature profile of islands for nucleation suppression at island edges although energy beams had a Gaussian profile.

For islands collared with terraces, lower temperatures were observed at groove edges. Thus single crystals grew along groove edges and expanded toward the center of islands resulting in a stable single crystal formation in grooves.

These observations pointed out that islands embedded in insulators with grooves and isolated single-scannings were the most suitable for better single crystallization because of the effective nucleation and stress suppressions at Si/SiO$_2$ interfaces. And further with a combination of proper island shapes/sizes and beam profiles, grain boundaries can be controlled and perfect single crystal islands can be obtained. Some seedings are, however, necessary for a complete orientation control since the preferred orientation was not perfect. More precise studies are necessary to clear ambiguities about the preferred crystal orientation and the growth habits.

Acknowledgements

Authors would like to express their thanks to S. Nakamura for his detailed TEM analyses of recrystallized specimens and their colleagues for IC processing in sample preparations. One of the authors, K. Kugimiya, would like to express his thanks to T. Takizawa and H. Ogawa of JEOL for their cooperation in fine modification of EB profiles.

This work was performed under the management of the R and D Association for Future Electron Devices as a part of the R and D Project of Basic Technology for Future Industries sponsored by Agency of Industrial Science and Technology, MITI.

REFERENCES

1) K. Kugimiya, G. Fuse, S. Akiyama, and A. Nishikawa: IEEE Electron Device Lett. **EDL-3** (1982) 270.
2) B. Y. Tsaur, J. C. C. Fan, R. L. Chapman, M. W. Geiss, D. J. Silversmith, and R. W. Mountain: IEEE Electron Device Lett. **EDL-3** (1982) 398.
3) Y. Ohmura, K. Shibata, T. Inoue, T. Yoshii, and Y. Horiike: IEDM Tech. Digest, No. 16-3 (1982) 429.
4) Y. Akasaka, T. Nishimura, and H. Nakata: 1983 Symposium on VLSI Tech., Maui (Sept. 14), 5-1 (1983).
5) T. Nishimura, A. Ishizu, and Y. Akasaka: Jpn. J. Appl. Phys. **22** (1983) 217.
6) T. Nishimura, Y. Akasaka, H. Nakata, A. Ishizu, and T. Matsumoto: *Proc. Soc. Information Display*, **23** (1982) 209.
7) J. P. Colinge and E. Demculin: IEEE Electron Device Lett. **EDL-2** (1981) 250.
8) G. T. Goele, E. W. Maby, D. J. Silversmith, R. W. Mountain, and D. A. Antoniadis: IEDM Tech. Dig. **24** (1981) 554.
9) T. Warabisako, M. Miyao, M. Ohkura, and T. Tokuyama: IEDM Tech. Dig. **16-4** (1982) 433.
10) T. J. Stultz and J. F. Gibbons: Appl. Phys. Lett. **39** (1981) 498.
11) S. Kawamura, J. Sakurai, M. Nakano, and M. Takagi: Appl. Phys. Lett. **40** (1982) 394.
12) T. Nishioka, N. Shinoda, and T. Ohmachi: *Proc. 43rd Meeting of Jpn. Soc. Appl. Phys.*, 28p-0-16 (1982) 373.
13) M. Aklufi, O. Csanadi, W. Dubbelday, and D. Silverberg: Electrochem. Meeting Ext. Abst., **82-1** (1982) 230.
14) J. P. Colinge, E. Demoulin, D. Bonahel, and G. Auvert: Appl. Phys. Lett. **41** (1982) 346.
15) J. F. Gibbons, K. F. Lee, T. J. Magee, J. Perg, and R. Ormond: Appl. Phys. Lett. **34** (1979) 831.
16) T. I. Kamins and P. A. Pianetta: IEEE Electron Device Lett. **EDL-1** (1980) 214
17) H. W. Lam, A. F. Tasch, Jr., and J. C. Holloway: IEEE Electron Device Lett. **EDL-1** (1980) 206.
18) D. K. Biegelsen, N. M. Johnson, D. J. Bartelink, and M. D. Moyer: Appl. Phys. Lett. **38** (1981) 150.
19) N. M. Johnson, D. K. Biegelsen, and M. D. Moyer: Laser and Electron Beam Solid Interactions and Material Processing, (1981) 463.
20) M. Tamura, H. Tamura, and T. Tokuyama: Jpn. J. Appl. Phys. **19** (1980) L23.

21) H. W. Lam, R. F. Pinizzotto, and A. F. Tasch, Jr.: ECS Ext. Abstr., Hollywood, Fla., Abstr. No. 481 (1980).
22) M. W. Geis, D. C. Flanders, D. A. Antoniadis, and H. I. Smith: IEDM Tech. Dig. (1979) 210.
23) K. Kugimiya, G. Fuse, S. Akiyama, and A. Nishikawa: Inst. Elect. Comm. Engr. Jpn. Tech. Rep., **SSD81-124** (1982) 67.
24) B. P. Von Herzen, T. I. Kamins, and C. I. Drowley: J. Electrochem. Soc. **128** (1981) 2695.
25) K. Kugimiya, E. Hirota, and Y. Bando: IEEE Trans. Magn., **MAG-10** (1974) 907.
26) S. Akiyama, N. Yoshii, S. Ogawa, S. Nakamura, and K. Kugimiya: The 14th Symp. on Ion Implant. and Submicron Fabrication, (1983) 97.
27) K. Kugimiya and S. Nakamura: To be presented at Proc. 44th Meeting of Jpn. Soc. Appl. Phys. (1983).
28) K. Kugimiya, G. Fuse, and K. Inoue: Jpn. J. Appl. Phys. **21** (1982) L19.
29) T. Morishita, M. Koba, and K. Awane: To be presented at IEEE SOS/SOI workshop (1983).
30) T. Nishioka, Y. Shinoda, and Y. Ohmachi: Appl. Phys. Lett. **43** (1983) 92.
31) R. P. Gale, J. C. C. Fan, R. L. Chapman, and H. J. Zeiger: *Mat. Res.Soc. Symp. Proc.* **V2** (1981) 439.
32) M. W. Geis, D. A. Antoniadis, D. J. Silversmith, R. W. Mountain, and H. I. Smith: Appl. Phys. Lett. **37** (1980) 454.
33) M. W. Geis, H. I. Smith, B. Y. Tsaur, J. C. C. Fan, D. J. Silversmith, and R. W. Mountain: J. Electrochem. Soc. **129** (1982) 2812.
34) T. Nishimura: private communication.
35) T. F. Ciszek, G. H. Schwuttke, and K. H. Yang: J. Cryst. Growth **50** (1980) 160.
36) L. Csepregi, E. F. Kennedy, J. W. Mayer, and T. W. Sigmon: J. Appl. Phys. **49** (1978) 3906.
37) J. Washburn: *Mat. Res. Soc. Symp. Proc.* **V2** (1981) 209.
38) K. Kugimiya, S. Akiyama, and S. Nakamura: Semiconductor Silicon, **81-5** (1981) 294.
39) T. Y. Tan, H. Foell, S. Mader, and W. Krakow: *Mat. Res. Soc, Symp. Proc.* **V2** (1981) 179.

Silicon-on-Insulator: Its Technology and Applications, edited by S. Furukawa, pp. 67–84.
© KTK Scientific Publishers, Tokyo, 1985.

RECRYSTALLIZATION OF SILICON ON INSULATOR WITH A HEAT-SINK STRUCTURE

S. KAWAMURA, N. SASAKI, T. IWAI, M. NAKANO, and M. TAKAGI

IC Development Division, Fujitsu Limited, Kawasaki 211, Japan

Abstract   Laser induced recrystallized silicon on insulator(SOI) is of technological interest as potential materials for 3-Dimensional integration. In fact, much progress has recently been made to enlarge the grain size of silicon films on insulator either by shaping the laser beam or by varying the power absorption of various regions of the substrate. However, most of these techniques leave a few residual grain boundaries in the recrystallized silicon mainly due to the failure of suppressing competitive nucleation. Therefore, how to remove the remaining grain boundaries has been the main focus in the recent SOI techniques. In this paper, a new method for obtaining complete single crystalline silicon films on SiO_2 with cw-Ar laser recrystallization is presented. The method utilizes the difference in thermal resistivity between the device regions with thin SiO_2 layers which act as a heat-sink and the surrounding regions with thick SiO_2 layers, thus controlling the nucleation and growth during resolidification process. 19-stage SOI/CMOS ring oscillators fabricated in the heat-sink structure have a propagation delay of 950psec per stage at a supply voltage of 7V. These results indicate that desired control of thermal profiling during resolidification process can be achieved by adjusting the structure of SOI.

1. Introduction

There has been a good deal of interest in the last few years in the formation of large area, large grain or single-crystalline silicon layers over amorphous insulating films by liquid phase recrystallization. Such layers can be used to fabricate three-dimensional integrated circuits or substrate-isolated high speed devices. These circuits are expected to offer the great advantages of reduced device parasitic capacitance, elimination of CMOS latch-up and improvement in radiation tolerance as well as reduced interconnection length and high packing density.

Melting of deposited films has been performed using a variety of energy sources such as cw-Ar laser, electron beams and strip heaters. In addition, techniques such as substrate patterning, substrate seeding and patterned islands have been used to increase the grain size of the recrystallized film.

The basic problem for the present SOI(Silicon-on-Insulator) technology is still how to increase the average grain size of the deposited films over insulator, although some devices have already been fabricated in the recrystallized films, showing good electrical characteristics. The increase in grain size and consequent decrease in grain boundary area results in improved silicon films, since grain boundaries are harmful to the electronic transport properties of the films. As a matter of fact, several techniques have been proposed to further increase the average grain size of the recrystallized film. The key to success is control of nucleation and growth during resolidification process. Therefore, the problem results in the interaction between an energy source and a substrate which is to be irradiataed.

In this paper, several techniques are presented to increase the grain size of scanned cw-Ar laser recrystallized silicon films over SiO_2 either by controlling the shape of the laser beam or by varying the power absorption(or dissipation) of various regions of the substrate. Changing the shape of the laser beam was carried out by using a doughnut-shaped beam instead of the usual Gaussian one, and continuous single-crystal silicon films of about $700\mu m$ over SiO_2 have been produced with the doughnut-shaped beam. Varying the power absorption(or dissipation) of various regions of the substrate was realized either with a structure having locally varied thickness of polysilicon or with a heat-sink structure in which device regions have thin SiO_2 layers acting as a heat-sink and the surrounding regions with thick SiO_2 layers, thus controlling the nucleation and growth during resolidification process. Complete single-crystallized device regions have been obtained by using the heat-sink structure, indicating that desired control of thermal profiling during resolidification process can be achieved by adjusting the structure of SOI, and the present heat-sink structure was found to offer ideal temperature distribution under laser irradiation.

2. *Experimental*

Figure 1 shows the schematic diagram of optical system for laser annealing. The two cylindrical lenses are used when an elliptical beam is required to widen the melt width. Basic laser annealing conditions are summarized in Table 1. For laser annealing the samples were mounted on a hot stage and the stage was driven in a serpentine scan by stepping motors with an average step of $15\mu m$. The beam was focused by a 48mm focal length lens.

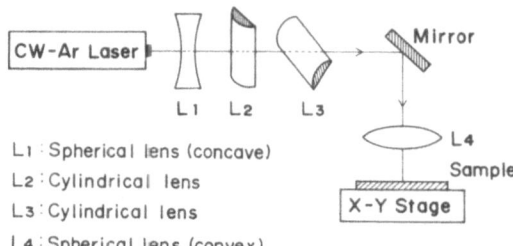

L1 : Spherical lens (concave)
L2 : Cylindrical lens
L3 : Cylindrical lens
L4 : Spherical lens (convex)

Experimental Setup

Fig. 1. Schematic diagram of optical system.

Table 1. Laser annealing conditions.

Laser : CW Ar$^+$ ion
Power : 8 W ∼ 12 W
Scan Speed : 5 cm/sec ∼ 10 cm/sec
Sub. Temp. : 400°C ∼ 500°C in air
Beam Shape : circular beam
          elliptical beam
Melt Width : 30 μm ∼ 50 μm
          ∼ 30 μm × ∼ 120 μm

3. Recrystallization of Si over SiO$_2$ by a Doughnut-Shaped Beam[1,2]

Controlling the shape of the beam is one of the techniques to obtain large grain silicon films over insulator as well as to understand the crystal growth mechanism under beam irradiation. In addition to the present doughnut-shaped beam technique, there have been some proposed so far such as using metal masks[3] or by using an oscillatory regrowth method.[4,5]

Basically, the grain size and shape of the recrystallized silicon are determined by the spacial distribution of the thermal profile and the following heat transfer at the trailing edge of the liquid-solid interface during solidification process. If the liquid-solid interface is concave, crystallites grow from the center of the molten zone to the outer edge, so that the competitive nucleation from the edge is suppressed, producing a large grain parallel to the scanning direction.

The polysilicon films used in this experiment were 400nm thick layers deposited by low pressure chemical vapor deposition on top of a 600nm thick insulating silicon oxide layer on a p-type (100) oriented Si substrate.

Stripe-patterned samples were also prepared to investigate the lateral epitaxial effect by the doughnut-shaped beam. Prior to laser recrystallization,
the samples were encapsulated with a 150nm silicon-nitride film and a ~ 1μm
PSG film to minimize mass transport of silicon as well as to decrease
optical reflectivity during laser irradiation. The 18W Spectra Physics Model
170 Ar ion laser used in this experiment has a 2m-long resonator composed of one flat mirror and one spherical mirror having a 6m radius for
the usual TEM_{00} mode. The spherical mirror was then replaced by one
that has a 4m radius to obtain a stable doughnut-shaped mode. The obtained doughnut-shaped mode is considered to be a linear combination
of TEM_{01}, and TEM_{10}, and often denoted by TEM_{01}^{*}. The typical mode
structure of Ar laser beam is schematically shown in Fig. 2. In the present
experiment, the output power of the doughnut-shaped beam is about 40%
of that of the usual Gaussian beam with TEM_{00} because of the energy
loss for higher mode structure.

Figure 3(a) shows the optical micrograph of the crystallization pattern resulting from a single laser scan with the doughnut-shaped beam
with an 11W output power, a 10cm/s scan speed after delineating the
grain boundaries by etching the sample with a diluted Wright etchant.
Here the liquid-solid interface line is supposed to be a saddle-shaped pro-

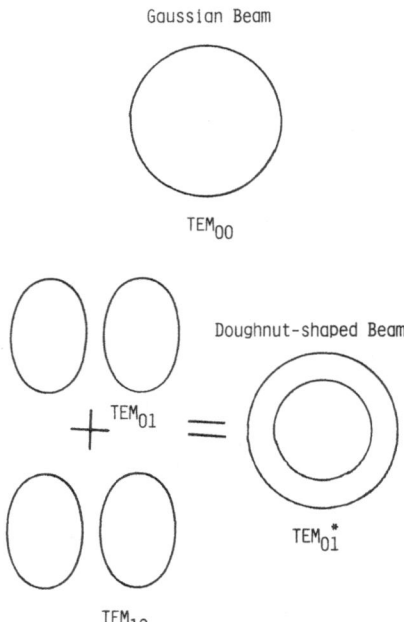

Fig. 2. Mode structure of laser beam.

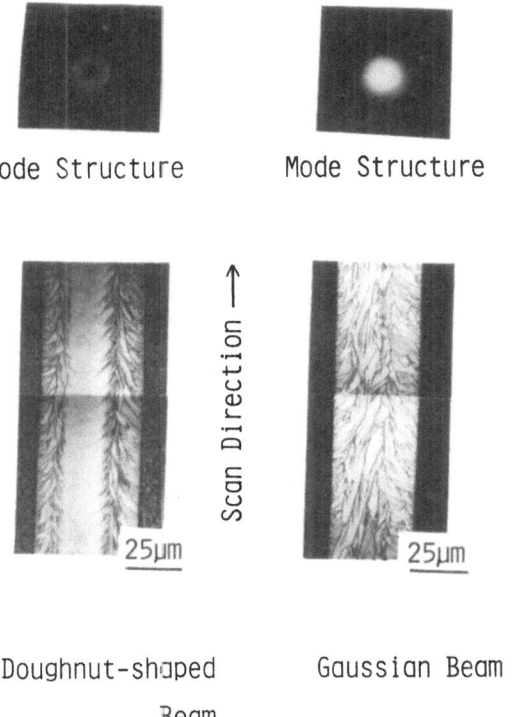

Fig. 3. Photomicrographs of the recrystallization pattern resulting from a single laser scan over SOI after Wright etching: (a) doughnut-shaped beam, (b) Gaussian beam. The photographs of the mode structure are also shown.

file as shown in Fig. 4(a) and a crystallite of polysilicon plays a role of a seed, though there is no bulk silicon seed in this case. Figure 3(b) shows the optical micrograph of the recrystallized pattern resulting from a single laser scan with a usual Gaussian intensity profile as a comparison. Here the nucleation of crystallites occurs at the trailing edge of the molten zone and competitive nucleation at the zone is not suppressed at all, yielding a large number of elongated grains as a result of the mutual blocking of the polycrystalline silicon growth from opposite zone edges. The liquid-solid interface line with consequent nucleation is schematically shown in Fig. 4(b). The photographs of the mode structure of the beam are also shown in Fig. 3.

The starting point of single crystal growth by the doughnut-shaped beam is shown in Fig. 5. It is clear from this picture that a grain which existd in the center of the molten zone played a role of a seed when the doughnut-shaped beam was applied to this material. The maximum length

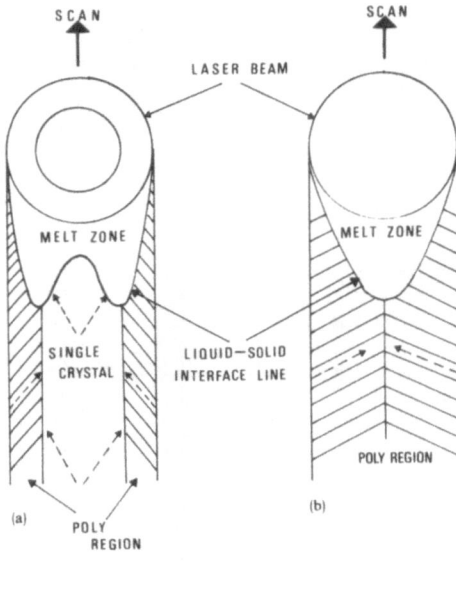

Fig. 4. Schematic illustrations of laser-induced molten spot and the liquid-solid interface line with consequent nucleation: (a) doughnut-shaped spot with concave trailing edge, (b) Gaussian spot with convex trailing edge.

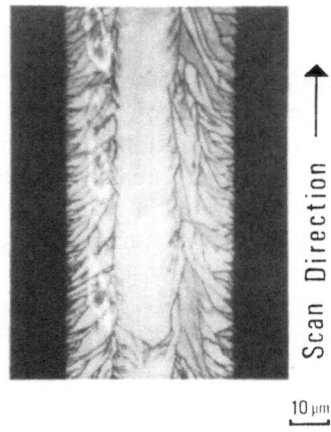

Doughnut – shaped beam

Fig. 5. Starting point of single crystal growth by doughnut-shaped beam.

of the single crystal obtained by the doughnut-shaped beam is about 700μm as shown in Fig. 6.

The problem thus far has been that the mode of the doughnut-shaped beam is not so stable as the Gaussian mode or TEM_{01} mode. The result of the recrystallization with a split beam(TEM_{01}) is shown in Fig. 7.

The sample with a patterned structure, or "lateral epitaxial structure", was also laser recrystallized by using the doughnut-shaped beam. Figure 8 shows the schematic cross section of the sample and the optical

Scan Direction ⟶

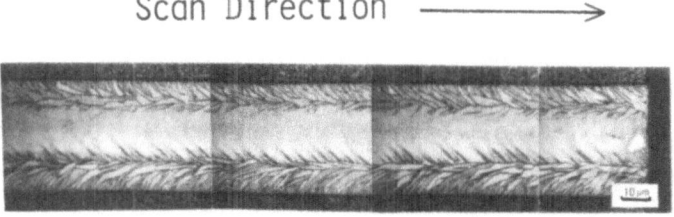

Fig. 6. Long single crystal silicon film produced by doughnut-shaped beam.

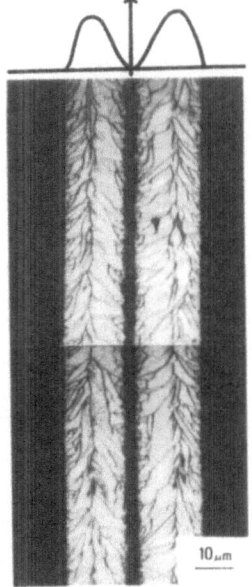

Split beam
TEM_{01}

Fig. 7. Recrystallized pattern with a split beam(TEM_{01} or TEM_{10}).

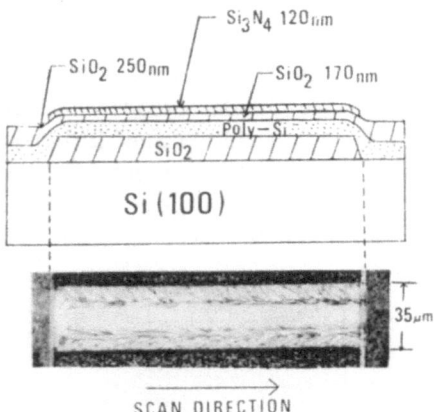

Fig. 8. Schematic cross section of the sample and the optical micrograph of the recrystallization pattern with doughnut-shaped beam.

micrograph of the crystallization pattern from a single laser scan with the doughnut-shaped beam with an output power of 12W, a scan speed of 10cm/s after delineating the grain boundaries. The photomicrograph clearly shows that the edge nucleation is suppressed and the longitudinal single crystal growth is seen along the beam scanning direction over SiO_2 region by lateral epitaxy from the bulk single-silicon seed. It should be noted that the doughnut-shaped beam can produce large grains regardless of the existence of the bulk silicon seed. Therefore, this technique can be applied to any kind of sample structures used for grain size improvement such as patterned or nonpatterned substrates. It does not seem that there is any fundamental limits on the distance of single crystal growth over SiO_2 that can be achieved by this procedure.

In conclusion, the grain size of laser-recrystallized SOI has been enormously increased by controlling the thermal profile of the molten silicon with a doughnut-shaped beam. It becomes clear that the most essential limiting parameter is the profile of the liquid-solid interface line, which has to be controlled so as to suppress edge nucleation during laser recrystallization.

4. Lateral Epitaxial Growth of Silicon on SiO_2 with Locally Varied Thickness of Polysilicon

In general, it is difficult to find a proper laser annealing condition when the recrystallization of SOI with a lateral epitaxial structure is performed. This is mainly because the thermal conductivity of Si is about 1.5w/cmdeg, while that of SiO_2 is about 0.014w/cmdeg. The two order

difference in thermal conductivity between Si and SiO_2 makes the window
in the laser power to induce liquid phase epitaxy both on Si seeds and
SiO_2 islands very narrow. In fact, when the laser power is high enough
to melt the polysilicon on the Si seed, the recrystallized silicon on the
SiO_2 island is thermally detached and left in Si droplets. However, when
the power is reduced to prevent the thermal detachment, the liquid phase
epitaxy on the Si seed does not occur sufficiently, producing a large number
of crystallites of various orientations over the SiO_2 island extended from
the oxide edge. These experimental results suggest that successful lateral
epitaxial growth of silicon layers over SiO_2 islands requires the dissipation
power of the recrystallized silicon to be roughly equal to each region of
the silicon seed or the SiO_2 island. To meet this requirement, a series
of laser induced recrystallization experiments were performed with samples
having locally varied thickness of polysilicon.

In this method, the thickness of polysilicon on SiO_2 islands is larger
than that on silicon substrates, thus making the thermal capacitance of
both regions roughly equal to realize desired lateral epitaxial growth.

The basic sample structure is schematically shown in Fig. 9. In this
figure, "a" is the thickness of polysilicon over SiO_2 islands, while "b"
that on silicon substrates. A series of experiments indicates that the best
result is obtained when $a = 400$nm, and $b = 200$nm. Figure 10 shows the
schematic cross section of the sample and the optical micrograph of
recrystallized silicon island over SiO_2 after etching the surface to delineate
grain boundaries. The picture shows that the silicon island is almost single-
crystallized with an output laser power of llW, a scan speed of 10cm/s,
and a 40μm beam width on the sample surface. Several silicon islands
were successfully single-recrystallized with this type of structure, as shown
in Fig. 11. The encapsulations used in this case compose of 60nm Si_3N_4
and 0.8μm PSG film. Figure 12 shows the SEM pictures of silicon island
at oxide edge before and after laser recrystallization. The SEM cross section of the recrystallized Si at the edge is also shown in Fig. 13. Smooth
continuous topography is an indication of sufficient lateral epitaxial growth
from the bulk silicon seed. In order to investigate the crystal orientation
of thus obtained single crystal silicon films over SiO_2, a micro focused

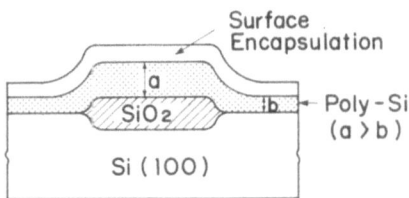

Fig. 9.   Schematic cross section of the sample having locally varied thickness of polysilicon.

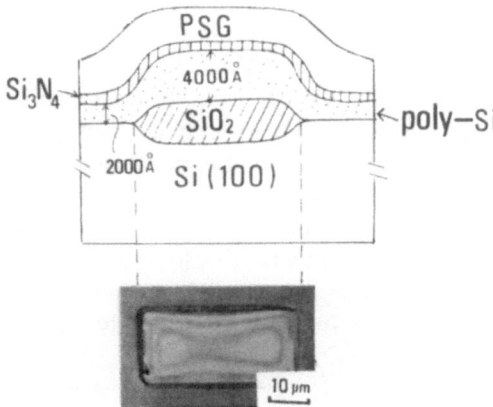

Fig. 10.   Schematic cross section and optical micrograph of recrystallized silicon island over SiO₂ after etching the surface to delineate grain boundaries.

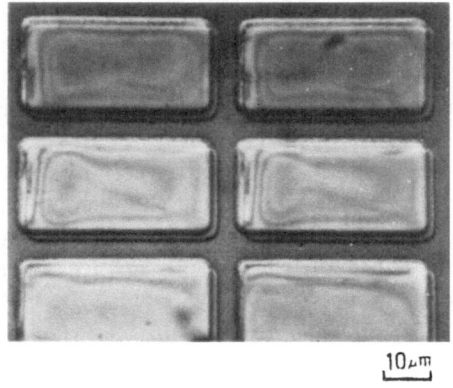

Fig. 11.   Single-recrystallized silicon islands over SiO_2.

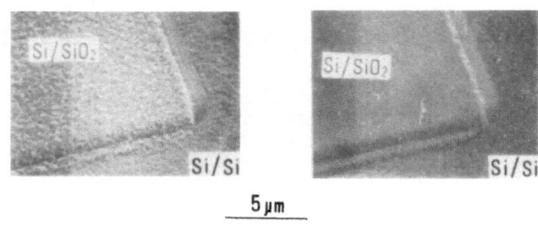

Before Laser Recrystallization   After Laser Recrystallization

Fig. 12.   SEM pictures of silicon island at oxide edge before and after laser recrystallization.

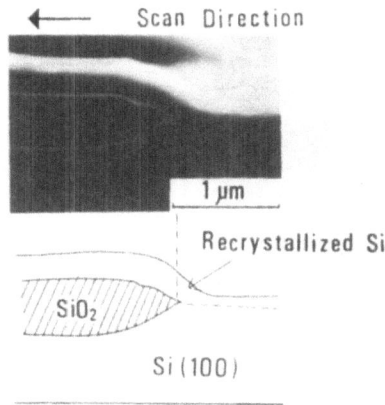

Fig. 13. SEM picture of cross section of the recrystallized SOI.

X-ray diffraction analysis was performed. The beam of the X-ray is focused to be about 30μm in diameter. Figure 14(a) shows the Laue Pattern obtained from the single crystallized region. The Laue Pattern is duplicated for a better view, as shown in Fig. 14(b), showing a four-fold rotational symmetry about an (100) crystal axis. Therefore, the obtained single crystal silicon film has (100) crystal orientation, confirming the lateral epitaxial growth from the bulk silicon seed.

5. Single-Crystallization of SOI with a Heat-Sink Structure

As mentioned above, much progress has been made in SOI technology to enlarge the grain size of as-deposited silicon films over insulator. However, most of these techniques leave a few residual grain boundaries in the laser recrystallized silicon mainly due to the failure of suppressing competitive nucleation of crystallites at the trailing edges of the molten zone or at the edges of the silicon islands. Therefore, how to remove the remaining grain boundaries has been the main focus in the recent SOI techniques.

Here, a new technique to obtain complete single crystalline silicon films over SiO_2 is presented. This technique utilizes the difference in thermal resistivity between the device regions having thin SiO_2 layers which act as a heat-sink during resolidification and the surrounding regions having thick SiO_2 layers. As the first step in sample preparation, a 600nm thick SiO_2 layer was thermally grown on single-crystal (100) Si substrates. Subsequently the oxide layer of the device region was selectively removed by approximately 400nm using a reactive ion etching technique, thus making a heat-sink structure as schematically shown in Fig. 15. 500nm

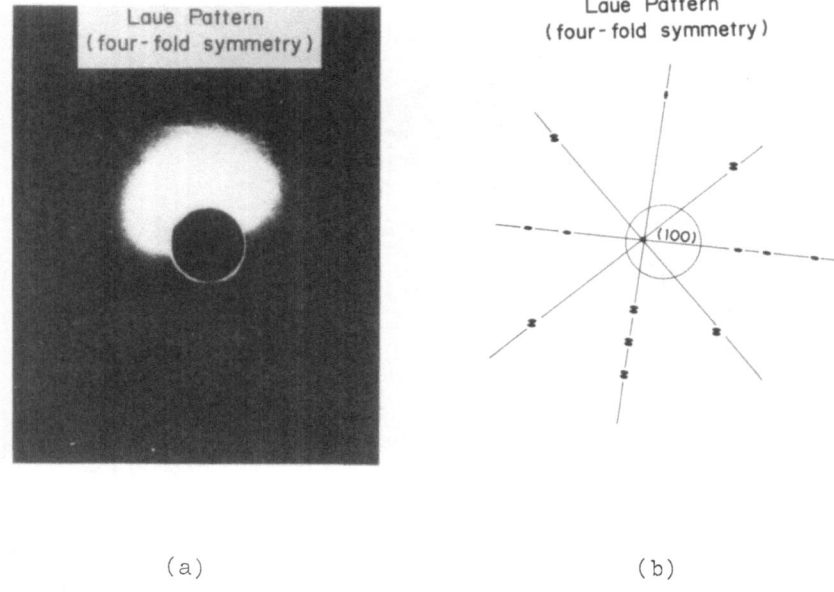

(a)   (b)

Fig. 14.   (a) Photograph of Laue Pattern. (b) Duplication of Laue Pattern, showing four-fold rotational symmetry around (100) crystal axis.

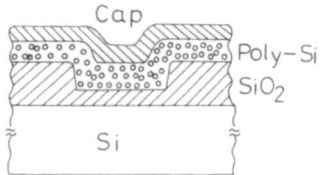

Fig. 15.   Schematic cross section of heat-sink structure.

polysilicon was then deposited by low pressure chemical vapor deposition, followed by the deposition of surface encapsulations such as a 120nm silicon-nitride film and a 0.8μm PSG film to minimize mass transport of molten silicon. The recrystallization of the deposited silicon films was performed with a multiline cw-Ar ion laser with a 9W output power and a 10cm/s scan speed.

In the experiments described here, the TEM_{00} output power has been formed into a collimated elliptical beam by two cylindrical lenses and focused to a spot $\sim 30\mu m \ X \sim 80\mu m$ which can cover the whole area of the device region in a single scan. The substrate temperature was kept at 480°C in air for all annealing experiments. Figure 16 shows the cross section and optical micrograph of the recrystallized silicon with a 9W output power after delineating the grain boundaries by etching the sample with a diluted Wright etchant. The photomicrograph in Fig. 16 shows that a complete single crystalline silicon film is produced in the dog-bone device region, while many grain boundaries are observed in the surrounding region. A series of experiments with different oxide thickness shows that the best result is obtained when the difference in thickness between the thin part and the thick part is about 350nm ~ 500nm with an output power of 8 ~ 10W. Figure 17 is a scanning electron microscopy photograph of the recrystallized region after etching the surface. It is clear from the picture

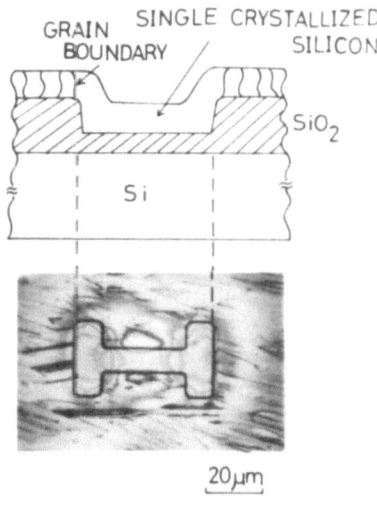

Fig. 16. Schematic cross section and optical micrograph of laser recrystallized silicon over SiO₂.

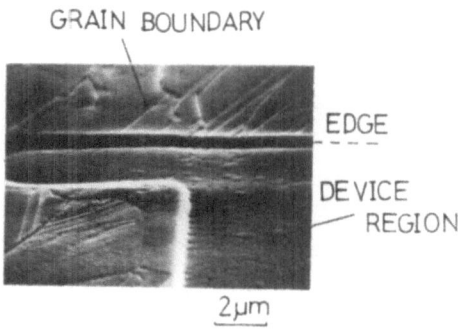

Fig. 17. SEM of laser recrystallized silicon film with a heat-sink structure.

that the growth of the grain boundaries is perfectly suppressed at the edge of the dog-bone device region, thus producing a complete single crystal film in that region.

The main idea of this technique is to control the temperature distribution of the device region during resolidification process by using heat-sink effect, thus realizing complete single-recrystallization at the cost of the surrounding silicon region where many grain boundaries are produced as usual. The spacial temperature distributions of two different types of SOI structure are shown in Fig. 18. For a conventional LOCOS island structure, the temperature profile is convex, since the heat dissipation rate around the island is larger than that in the center of the island, as shown in Fig. 18(a), so that the nucleation of crystallites occurs at the edge of the island, yielding a large number of grains. On the other hand, for the heat-sink structure, the concave temperature profile, as shown in Fig. 18(b), prevents grain boundaries nucleated around the island edges from propagating towards the island center. It should be noted that, in the heat-sink structure, the silicon film is not defined into device islands before laser recrystallization, which offers an ideal temperature distribution during resolidification process. As a matter of fact, it was found that some grain boundaries were produced or agglomeration was observed in the device region when the silicon islands are already defined before laser recrystallization. Another advantage of the heat-sink structure is that, even if the laser power is quite high, agglomeration occurs only in the surrounding region and not in the device region, as shown in Fig. 19. In the present experiment, with an output power of over 12W, agglomeration occurs in the surrounding region, but never in the dog-bone device region.

Micro focused X-ray diffraction analysis indicates that (100) texture is dominant in the single-crystallized area, whereas mixed orientations in the surrounding region.

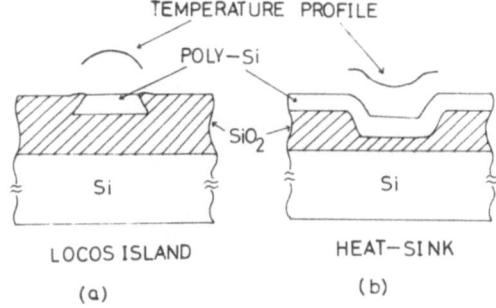

Fig. 18.   Schematic cross section and illustration of temperature profile during resolidification: (a) conventional LOCOS island structure, (b) heat-sink structure.

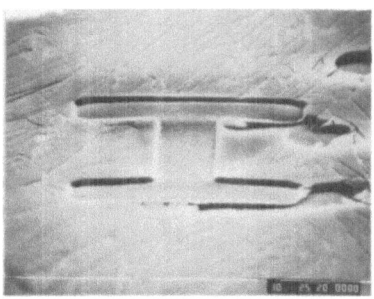

10 μm

Fig. 19. SEM of laser recrystallized silicon film with a heat-sink structure viewed at a
near glancing angle. Silicon detachment is observed outside device region.

In conclusion, using a heat-sink structure in order to adjust the thermal resistivity of the device region and the peripheral region, a complete single crystal silicon island has been obtained. The result indicates that desired control of thermal profiling during resolidification process can be achieved by adjusting the structure of SOI, and the present heat-sink structure was found to offer ideal temperature distribution under laser irradiation with edges of the device region hotter than the center, so that the heat sinks sufficiently to initiate recrystallization there.

6. Fabrication of SOI/CMOS Ring Oscillator with a Heat-Sink Structure

After recrystallization of the silicon films on SiO_2 with the heat-sink structure, CMOS invertors were fabricated to investigate the crystalline quality of the recrystallized device region. 19-stage CMOS ring oscillators with 4μm channel length have also been fabricated. Figure 20 shows the

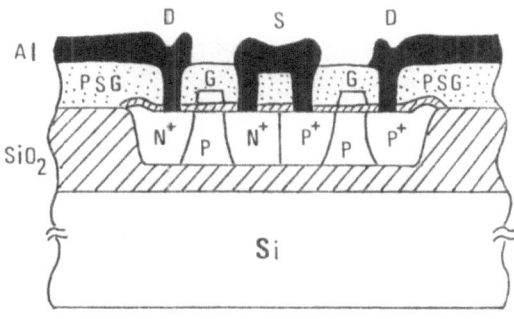

Fig. 20. Schematic cross section of SOI/CMOSFET with a heat-sink structure.

schematic cross section of the fabricated CMOSFET with a heat-sink structure. A photomicrograh of a plan view of the CMOSFET is also shown
in Fig. 21. Figure 22 shows *I-V* characteristics for typical p- and n-channel

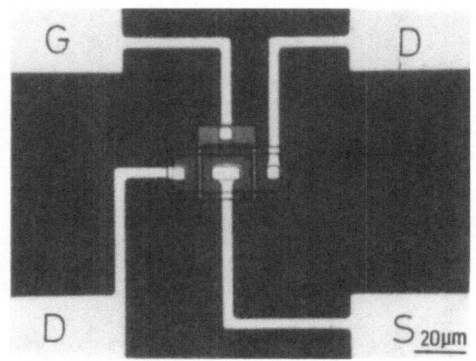

Fig. 21.   Plan view of CMOSFET.

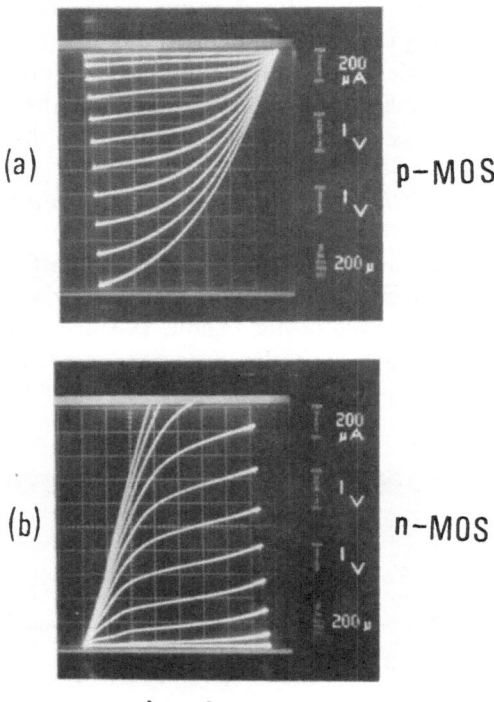

Fig. 22.   *I-V* characteristics of typical transistors with channel length of 4μm: (a) p-channel
device, (b) n-channel device.

transistors with $W = 18\mu m$ and $L = 4\mu m$. From the I-V characteristics, the mobility of the holes in the p-channel device is calculated to be 210cm^2/Vs, while that of the electrons in the n-channel device is 500cm^2/Vs. The threshold voltage is 1.0V for the n-channel device and -0.8V for the p-channel device. These threshold voltages can be adjusted simultaneously by a suitable channel-implant doping. A microscopic view of the 19-stage ring oscillator fabricated in the heat-sink structure is shown in Fig. 23. Figure 24 is the output waveform of the ring oscillator with an output buffer stage and with $4\mu m$ channel length. The circuit is observed to oscillate with a supply voltage of 6V–12V and the minimum propagation delay was calculated to be 950psec per stage at a supply voltage $V_{DD} = 7V$.

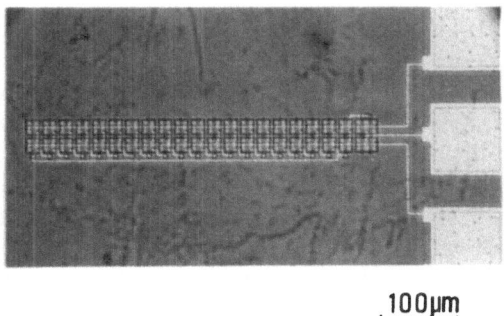

100μm

Fig. 23. Plan view of 19-stage SOI/CMOS ring oscillator.

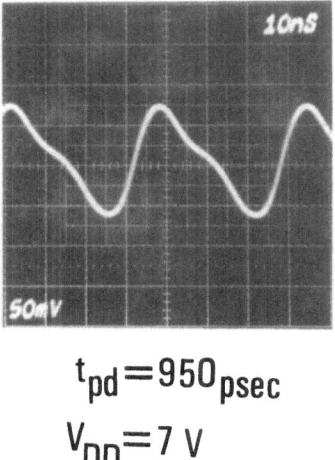

$$t_{pd} = 950 psec$$

$$V_{DD} = 7 V$$

Fig. 24. Output waveform of the ring oscillator.

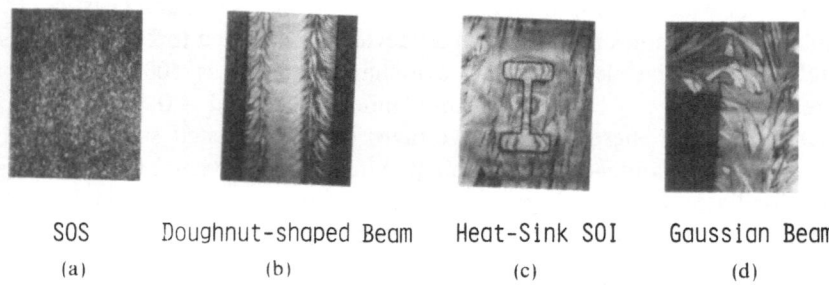

SOS   Doughnut-shaped Beam   Heat-Sink SOI   Gaussian Beam
(a)   (b)   (c)   (d)

Fig. 25. Photomicrographs of recrystallized surface after Wright etching: (a) SOS, (b) SOI by doughnut-shaped beam, (c) SOI with heat-sink structure, (d) SOI by Gaussian beam.

7. Comparison of SOI with SOS by Wright Etching

Large grain crystallization of polysilicon layers on insulating materials by beam processing offers the potential of producing device quality SOI structures for dielectric isolation as an alternative to SOS(Silicon-on-Sapphire). As a matter of fact, the devices fabricated in the beam recrystallized silicon films on SiO_2 show better electrical characteristics than those fabricated in SOS.[6] Here the crystalline quality of SOI is compared with that of SOS by using Wright etching techniques. Figure 25 shows the etched-surface of (a)SOS, (b)SOI by a doughnut-shaped beam, (c)SOI with a heat-sink structure, and (d)SOI by a Gaussian beam. It is clear from these pictures that the quality of SOI is far better than that of SOS as long as the etching results are concerned. These results should encourage the further development of SOI technology with the potential for achieving higher speed than SOS in VLSI circuits.

REFERENCES

1) S. Kawamura, J.Sakurai, M. Nakano, and M. Takagi: Appl. Phys. Lett. **40** (1982) 394.
2) S. Kawamura, J. Sakurai, and M. Nakano: *161st ECS Meeting*, May (1982) No. 150, P.243.
3) T. J. Stultz and J. F. Gibbons: Appl. Phys. Lett. **39** (1981) 498.
4) G. K. Celler, L.E. Trimble, K. K. Ng, H. J. Leamy, and H. Baumgart: Appl. Phys. Lett. **40** (1982) 1043.
5) H. Ishiwara, M. Nakano, H. Yamamoto, and S. Furukawa: *Proc. 14th Conf. Solid State Devices, Tokyo, 1982*, Jpn. J. Appl. Phys. **22** (1983) Suppl. 22-1.
6) M. W. Geis, H. I. Smith, B-Y. Tsaur, and J. C. C. Fan: Appl. Phys. Lett. **40** (1982) 158.

Silicon-on-Insulator: Its Technology and Applications, edited by S. Furukawa, pp. 85–97.
© KTK Scientific Publishers, Tokyo, 1985.

RECRYSTALLIZATION OF POLYCRYSTALLINE Si OVER SiO$_2$ THROUGH STRIP ELECTRON-BEAM IRRADIATION

Y. HAYAFUJI,[1] T. YANADA,[1] S. USUI,[1] S. KAWADO,[1] A. SHIBATA,[1] N. WATANABE,[1] M. KIKUCHI,[1] H. HAYASHI,[2] and K. E. WILLIAMS[3]

[1]Sony Corporation Research Center, 174 Fujitsuka-cho, Hodogaya-ku, Yokohama, Japan
[2]Semiconductor Division, Sony Corporation, Asahi, Atsugi 243, Japan
[3]Energy Science Inc., 8 Gill Street, Woburn, Mass. 01810, U.S.A.

Abstract Polycrystalline silicon on SiO$_2$ islands was successfully recrystallized by rapid scanning with a strip electron beam. The electron-beam was focussed on the samples in a strip measuring 3 cm × 60 μm, with a voltage of 10 kV and with peak current densities up to 70 A/cm^2. Samples with a poly-Si/(001)Si, poly-Si/SiO$_2$/(001)Si and poly-Si/striped SiO$_2$/(001)Si structure were prepared.

In the poly-Si/striped SiO$_2$/(001)Si structure, lateral epitaxial growth in the scanned direction extended as much as 70 μm from the seeded area at the edge of the 100-μm-wide stripe. The regrowth speed was estimated to be about 200 cm/sec. TEM analysis showed that the recrystallized area on the SiO$_2$ stripe was (001) single-crystalline film which included small-angle boundaries. The surface appearance depended on the regrowth direction and the ⟨110⟩ direction was found to be the preferential direction. Encapsulation of the samples with silicon nitride film effectively smoothed the recrystallized surface, although small-angle boundaries still were in evidence.

1. Introduction

Recrystallization of a thin single-crystalline silicon film on an insulating material, the so-called SOI (silicon-on-insulator) structure, is the basis of a new semiconductor technology which may lead to the development of a semiconductor device composed of multiple active layers, a 3D IC.

A common technique to form SOI thin single-crystalline film is to melt (by scanning a high energy density spot laser or spot electron beam[1–5] or by moving a strip heater) and regrow polycrystalline silicon (poly-Si) film which has been deposited on SiO$_2$ film by chemical vapor deposition.

The scanning pattern of the spot beam is serpentine and it is difficult to obtain a large single-crystalline silicon film by this technique as a boundary exists between adjacent scanning traces. Systems utilizing an arc strip lamp[6] or a strip graphite heater,[7,8] however, are successful in producing a large area single-crystalline silicon film. The use of a graphite heater realizes SOI films of good crystallinity but this process is limited in its application to mono-layer devices, as it is necessary to heat the substrate to well above 1000°C so that the radiant energy from the graphite strip heater can melt the poly-Si film. Knapp et al.[9,10] have suggested that a strip-shaped laser or electron beam might be capable of producing uniform single-crystalline SOI at relatively low substrate temperatures but to date this technique has not been successfully applied to produce SOI of a large area. In this paper, we describe laterally seeded regrowth of a large area SOI film using a strip electron beam of a high energy density with a width of less than 100 μm. The regrowth speed and the preferential direction of laterally seeded regrowth are also presented. A part of the present study has been previously reported.[11]

2. Experimental

The samples used in this study were boron-doped (001) Si CZ wafers with resistivity ranging from 8 to 10 Ω-cm. The diameter and thickness of the wafers were 76 mm and approximately 400 μm, respectively. The front surface was mechano-chemically polished, and the back surface was finished by chemical etching. The wafers were cleaned using a conventional chemical solution and rinsed in de-ionized water before deposition of poly-Si.

The samples were divided into three groups acording to the structure of the poly-Si layer, as illustrated in Fig. 1. Samples in the first group (Fig. 1a) had poly-Si film 0.5 μm thick over the entire surface of the wafer. An insulating film was not formed between the poly-Si and the substrate. Poly-Si was deposited by LPCVD at 650°C using a mixture of SiH_4 and N_2. The poly-Si film of the second group of samples was deposited to a thickness of 0.5 μm over the 0.5 μm thermal SiO_2 film grown at 1000°C in a dry O_2 atmosphere for 2 hours (Fig. 1b). In the third group of samples (Fig. 1c), a poly-Si film of 0.5 μm again, was deposited over SiO_2 islands formed by conventional photolithographic process. The SiO_2 film was the same as in the second group of samples. The samples in the third group were used to investigate laterally seeded recrystallization. The width and the separation between adjacent islands were 100 μm. The islands were 2 cm long. The areas betwen two adjacent islands, from which the SiO_2 film had been removed, were used as seeding areas for recrystallization. The long edge of an island was in the [010]

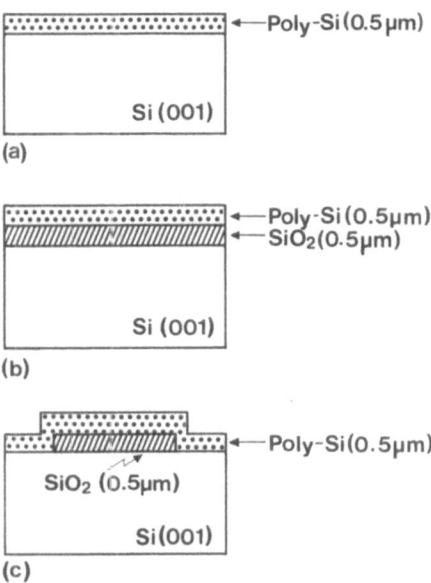

Fig. 1.   Cross-sectional view of samples used in electron-beam irradiation experiments. (a) Poly-Si/(001)Si, (b) Poly-Si/SiO₂/(001)Si, (c) Poly-Si/striped SiO₂/(001)Si.

direction and the short edge in the [100] direction. Some of the samples in the third group were covered with a 400 Å silicon nitride film deposited by LPCVD using a mixture of SiH_2Cl_2 and NH_3 at 750°C. These samples were used to study the effect on recrystallization of encapsulation by the nitride film.

The poly-Si film was recrystallized by irradiating a strip electron beam. The strip electron-beam system was operated at an acceleration voltage of 10 kV. The electron range of 10 keV electrons into Si is about 0.4 μm. An electron beam generated from a cathode was focussed onto a strip approximately 60 μm wide. The length of the strip electron beam was 3 cm on the wafer surface. The beam current was 1.0 to 1.3 A for a beam duration of 0.1 to 2 msec. The current density on the surface of the wafer was estimated to be 70 A/cm², which would mean an energy density of 0.7 MW/cm². The wafer was moved so that the stationary strip electron beam crossed the oxide islands at a speed of up to 500 cm/sec. A heater was provided below the sample holder to preheat the wafer to temperatures up to 900°C prior to electron beam irradiation. Electron-beam irradiation for 1 msec melted a 4.2 mm length of poly-Si film when the sample was moved at a speed of 420 cm/sec.

The as-deposited poly-Si films were investigated by transmission elec-

tron microscopy (TEM) and by transmission electron diffraction (TED).
The morphology of the recrystallized Si film was examined by an optical
microscope and also by Talystep measurement using a stylus with an area
of 0.1 μm × 2.5 μm. The recrystallized Si film was characterized
crystallographically by TEM. Samples for TEM analysis were prepared
by thinning the wafers to less than 1 μm thick by chemical etching.

3.   *Experimental Results*

The grain size and orientation of the deposited poly-Si film were checked prior to recrystallization by electron-beam irradiation. Figure 2a gives
a dark-field TEM image of the deposited poly-Si film and Figure 2b shows
the TED pattern of the area corresponding to Fig. 2a. These figures show

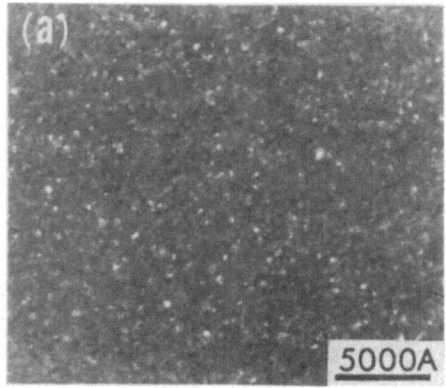

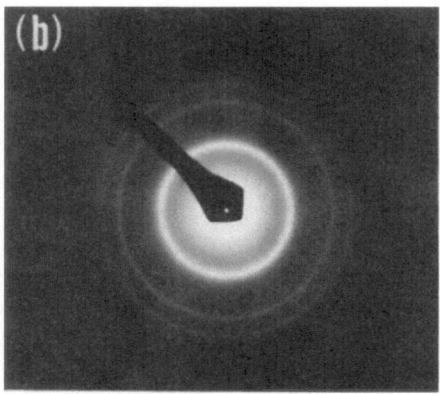

Fig. 2.   (a) TEM dark-field image of as-deposited polycrystalline film; (b) TED pattern
of the same area shown in Fig. (a).

that poly-Si film deposited on a substrate has random crystallographic orientation and that its grain size is about 200 Å.

In order to determine the beam width of the present system, we melted various stationary samples and observed the width of the melted area. Figures 3a, 3b and 3c show examples of the melt region in samples irradiated under various conditions. It is apparent that the melted region is 60 μm wide in all three examples. We consider this value to be the effective beam width. The energy distribution across the beam width is expected to be quasi-Gaussian but the actual energy distribution of the electron beam was not determined.

Figure 4 shows the current level as a function of the sample speed

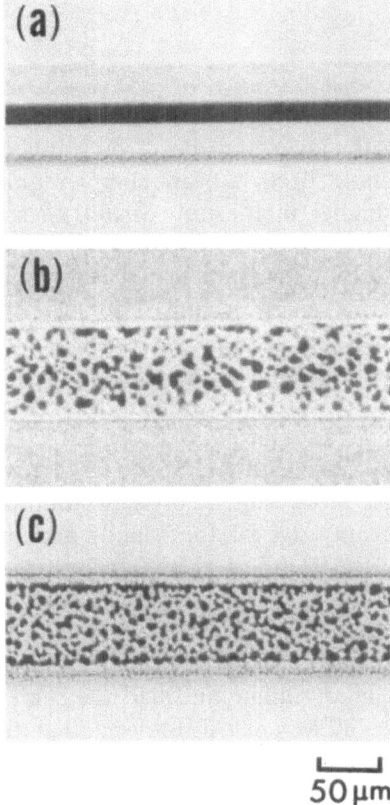

50 μm

Fig. 3. Optical micrographs showing recrystallized areas irradiated in a stationary state. (a) Single Si wafer irradiated by 200 μsec electron-beam pulse at 1.3 A and 10 kV. (b) Wafer with a poly-Si/(001)Si structure irradiated by 180 μsec pulse at 1.3 A and 10 kV. (c) Wafer with a poly-Si/SiO₂ structure irradiatd by 160 μsec pulse at 1.2 and 10 kV

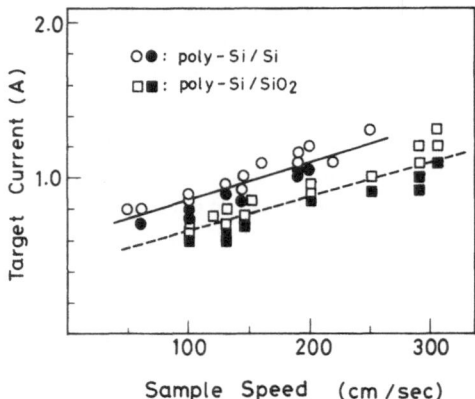

Fig. 4.   Threshold target current required to melt poly-Si films at various sample speeds in poly-Si/SiO₂ structure.

required to melt the poly-Si film on Si substrates both with and without a thermally grown SiO₂ layer at a beam duration of 1 msec at 10 kV. The data give a threshold electron-beam energy required to melt the poly-Si film about 20% higher in the poly-Si/(001)Si structure than that in the poly-Si/SiO₂/(001)Si structure. The difference in heat dissipation into the substrate in these two structures can be attributed to the difference in the threshold melting energy required to melt the poly-Si film.

Figure 5a shows an optical micrograph of a laterally seeded regrowth film on an oxide island 100 μm wide. The surface profile measured by Talystep is shown in Fig. 5b. The sample was preheated to about 550°C before electron-beam irradiation with a beam of 1.3 A at 10 kV for 1 msec and a sample speed of 300 cm/sec. The arrow in the figure shows the direction of sample movement. Recrystallization proceeded in the [100] direction from the front edge of the islands and in the [Ī00] direction from the rear edge of the island. A ridge 0.5 μm high was formed between the middle of the island and the rear edge. In the area regrown from the front edge, a closed-loop pattern is observed. The TEM image and TED patterns of the regrown area are shown in Fig. 6. The regrown film was found to be single-crystalline and had the same (001) orientation as that of the substrate. TEM analysis showed that the closed loops corresponded to small-angle boundaries and the ridge corresponded to large-angle boundaries. Misorientation across a small-angle boundary was less than 1 degree, while misorientation across a large-angle boundary was in a range of 5 to 10 degrees.

Laterally seeded regrowth was also attempted in a different crystallographic orientation in order to study the regrowth orientation

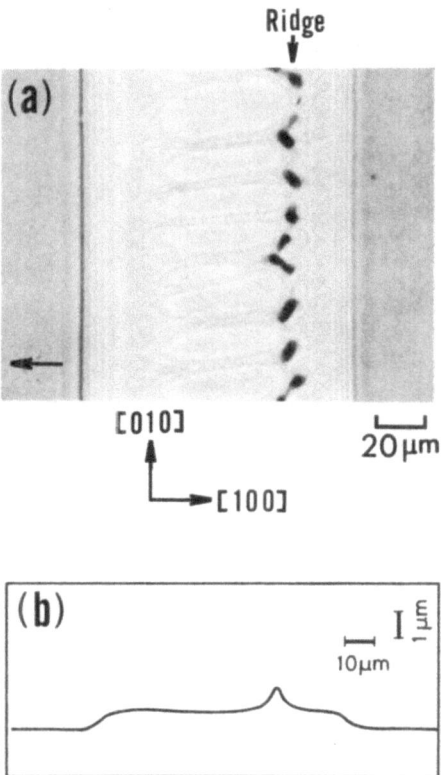

Fig. 5. (a) Optical micrograph showing lateral-seeding epitaxial growth over a striped SiO$_2$ film 100 μm wide. Arrow indicates the direction of sample movement. (b) Talystep profile of the regrown area measured in the regrowth direction. The direction of the measurement is perpendicular to the stripe edge.

preferred. Figure 7 shows an optical micrograph of a regrown film where the regrowth proceeded in the [110] direction from the front edge of the island and in the opposite direction from the rear edge. In contrast to the zig-zag ridge observed in Fig. 5, a straight ridge was formed in this sample. Another characteristic of regrowth in a [110] direction is the many lines along the edge of the island. Also, the closed loops observed in regrowth in the [100] direction were not formed.

The effect of encapsulation on laterally seeded recrystallization was investigated using samples covered with a 500-Å-thick silicon nitride film over a poly-Si film deposited on thermal oxide. Figure 8a is an optical micrographs of an encapsulated sample irradiated under the same conditions as in Fig. 5. Figure 8b is the Talystep profile. Generation of closed

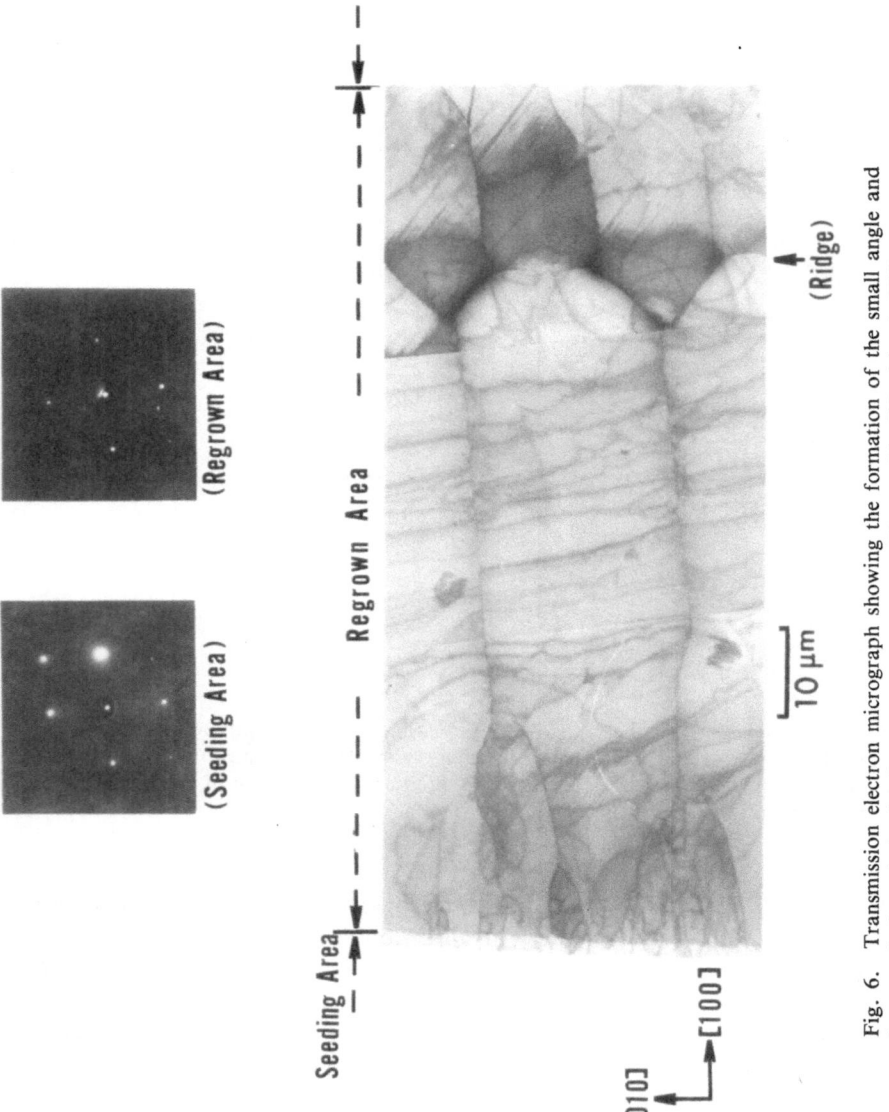

Fig. 6. Transmission electron micrograph showing the formation of the small angle and large-angle boundaries in the regrown area on a 100-μm-wide SiO₂ island. TED patterns of the seeding area and regrown area showing the same (001) net pattern.

Ridge

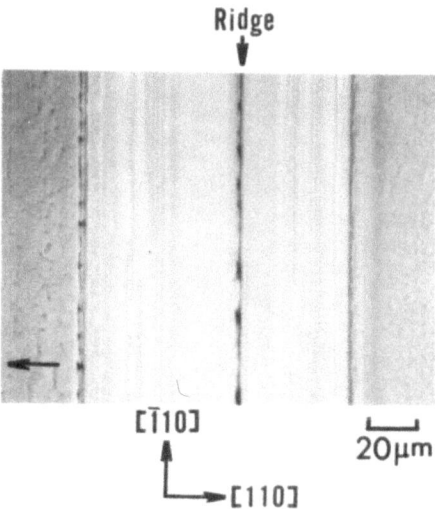

[$\bar{1}$10]

[110]

20μm

Fig. 7. Optical micrograph of the recrystallized area on a striped SiO$_2$ film 100 μm wide. The direction of the regrowth is [110].

loops was suppressed in this sample. A series of isolated protrusions was found in contrast to the ridge observed in the regrown area of non-encapsulated wafer. Talystep measurement gave a height of these protrusions of about 0.1 μm. The encapsulating film made the surface of the crystallized film more smooth. TEM observation of the sample showed that there still existed small-angle grain boundaries in the regrown area. This suggests that there was no significant improvement in the crystallinity of the regrown layer.

4. Discussion

Irradiation of a stationary strip electron beam can melt the surface layer of a silicon wafer, forming a ridge at the middle of the melted area as is shown in Fig. 3a. The ridge may be formed in the following way. When the melt is very shallow and the heat flow in recrystallization is perpendicular to the surface, it is conceivable that no ridge can be formed. The regrowth velocity, however, has a component parallel to the sample surface as well as a component perpendicular to the surface. In view of the fact that the meniscus angle at the solid-liquid interface in silicon is greater than 10 degrees and the density of liquid silicon is greater than that of solid silicon, the melted silicon is squeezed towards the center of the melted region to form a ridge. The ridge is formed when a sample with a poly-Si/striped SiO$_2$/(001)Si structure is moved with respect to a

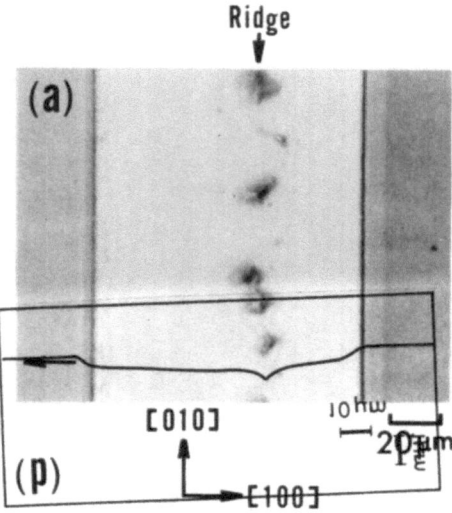

Fig. 8.   Optical micrograph (a) and Talystep profile (b) of an encapsulated sample irradiated under the same conditions as in Fig. 4.

stationary strip electron beam. As is shown in Fig. 5, a ridge is formed when two melt fronts proceeding from both SiO$_2$ edges collide. The ridge, however, is not at the middle of the island but somewhere between the middle and the rear edge of the island.

The propagation speed of the regrowth front during laterally seeded regrowth of poly-Si film on an SiO$_2$ island was roughly estimated from the sample speed and the position of the ridge. Refer to Fig. 9, which is schematic of the temperature distribution in a sample irradiatd by an electron beam. We assume that the average speed, V_r, of the regrowth front moving from the front edge is equal to the average speed of regrowth front moving from the rear edge. We also assume that the interval t_o which the melted area remains liquid is the same in both seeding areas adjacent to an SiO$_2$ island. When electron-beam irradiation starts from the front SiO$_2$ edge, $x = 0$, at time $t = 0$, the distance L_1 from the front

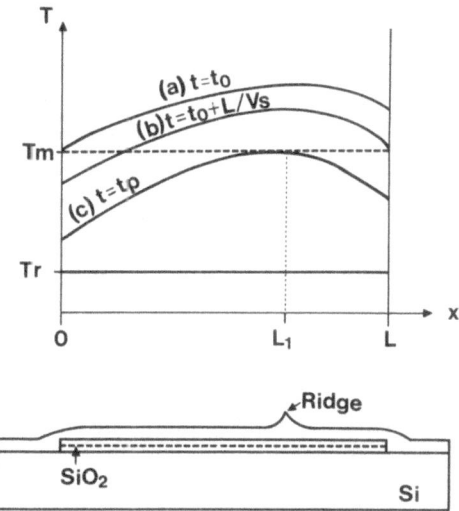

Fig. 9. Schematic illustration used to derive regrowth speed. T_m is the melting point of Si and L_1 is the position of the ridge. Temperature distribution varies with time in the sequence (a), (b) and (c).

SiO$_2$ edge to the ridge can be expressed as

$$L_1 = V_r(t_p - t_0) \tag{1}$$

where t_p is the time at which a ridge is formed. The distance between a ridge and the rear edge is given by

$$L - L_1 = V_r\{t_p - (t_0 + L/V_s)\} \tag{2}$$

where L is the width of an island and V_s is the speed of the sample relative to the stationary electron beam. From the above equations, we can obtain the average regrowth speed,

$$V_r = V_s(2L_1 - L)/L. \tag{3}$$

In a typical case, V_s, L and L_1 were 300 cm/sec, 100 μm and 75 μm, respectively, and the average regrowth speed is 150 cm/sec. Calculations give a regrowth speed in the range of 150 to 250 cm/sec on 0.5 μm-thick SiO$_2$ islands 100 μm wide. These values are much higher than the speed of 0.01 to 0.7 cm/sec for growing Si ribbons by various methods,[12-15] but are comparable to the speed of the explosive crystallization of amorphous Si film on an insulating substrate, which has been estimated to be as high as 1000 cm/sec.[16,17] A similar analysis of recrystallization speed of Si on SiO_2 was recently made by Knapp and Picraux.[18]

TEM analysis confirmed that the laterally seeded regrowth film on an SiO_2 layer extended as much as 70 μm. The distance of recrystallized film from the SiO_2 edge is limited by the formation of the ridge, so that the use of a sample with a single seeding area at the front edge of the SiO_2 island is necessary in order to extend the recrystallized film over 70 μm. The formation of small-angle boundaries may be caused by thermal stress induced by the difference in the thermal expansion coefficients of Si and SiO_2. The closed-loop pattern observed on the recrystallized surface of the sample in Fig. 5 reflects to the change in the thickness of regrown film at the small-angle boundaries. It is interesting to note that the surface appearance of the sample in Fig. 7 differs from that in Fig. 5. When the regrowth direction is [001], the ridge runs zigzag. When the regrowth direction is [110], however, the ridge runs straight and many straight lines along the SiO_2 edge can be observed in the recrystallized area. This means that the preferential direction of laterally seeded regrowth over SiO_2 islands is [110].

5. Summary

Single-crystal films were obtained on SiO_2 islands 100 μm wide and 2 cm long by laterally seeded regrowth of poly-Si films using a strip electron beam measuring 3 cm $\times$ 60 μm with an energy density of 0.7 MV/cm^2. The recrystallization speed of lateral regrowth of the poly-Si film over SiO_2 islands was estimated from the position of the ridge, the width of the SiO_2 island and the sample speed, to be about 200 cm/sec. This value is much higher than the regrowth speed observed in other regrowth methods. The surface appearance depended on the regrowth direction and the ⟨110⟩ direction was found to to be preferential.

Acknowledgements
The authors wish to thank T. Tanigaki, S. Kumagai, K. Hall, T. Avnery, M. Fletcher and T. Carrao for their assistance in the present work.

REFERENCES

1) A. Gat, L. Gerzberg, J. F.Gibbons, T. J. Magee, J. Peng, and J. D. Hong: Appl. Phys. Lett. **33** (1978) 775.
2) M. W. Geis, D. C. Flaunders, and H. I. Smith: Appl. Phys. Lett. **35** (1979) 71.
3) M. Tamura, H. Tamura, and T. Tokuyama: Jpn. J. Appl. Phys. **19** (1980) 123.
4) K. Shibata, T. Inoue, T. Takigawa, and S. Yoshii: Appl. Phys. Lett. **39** (1981) 645.
5) T. I. Kamins and B. P. Von Horzen: IEEE Electron Devices Lett. **EDL-2** (1981) 313.
6) T. J. Stultz and J. F. Gibbons: Appl. Phys. Lett. **41** (1982) 825.

7) J. C. C. Fan, M. W. Geis, and B. Y. Tsaur: Appl. Phys. Lett. **38** (1981) 365.

8) R. F. Pinizzotto, H. W. Lam, and B. L. Vaandrager: Appl. Phys. Lett. **40** (1982) 388.

9) J. A. Knapp and S. T. Picraux: Appl. Phys. Lett. **38** (1981) 873.

10) J. A. Knapp and S. T. Picraux: J. Appl. Phys. **53** (1982) 1492.

11) Y. Hayafuji, T. Yanada, S. Usui, S. Kawado, A. Shibata, N. Watanabe, M. Kikuchi, and K. E. William: 163rd ECS Mtg. San Francisco, CA. (1983).

12) J. P. Kalejs, B. H. Mackintosh, and T. Surek: J. Cryst, Growth **50** (1980) 175.

13) A. Baghdadi and R. W. Gurtler: J. Cryst. Growth **50** (1980) 236.

14) B. Kudo: J. Cyst. Growth **50** (1980) 247.

15) J. D. Zook, B. G. Koepke, B. L. Grung, and M. H. Leipold: J. Cryst. Growth **50** (1980) 260.

16) M. Kikuchi, A. Matsudo, T. Kurosu, A. Mineo, and M. J. Calaman: Solid State Commun. **14** (1974) 731.

17) G. Aurvert, B. Bensahel, A. Georges, V. T. Nguyen, P. Henoc, F. Morin, and P. Coissard: Appl. Phys. Lett. **38** (1981) 613.

18) J. A. Knapp and S. T. Picraux: 1982 MRS Symposium, Nov., Boston Mass.

CHAPTER 2 : ZONE MELTING RECRYSTALLIZATION

Silicon-on-Insulator: Its Technology and Applications, edited by S. Furukawa, pp. 101–128.
© KTK Scientific Publishers, Tokyo, 1985.

ZONE-MELTING RECRYSTALLIZATION OF Si FILMS ON SiO$_2$

B-Y. TSAUR

Lincoln Laboratory, Massachusetts Institute of Technology, Lexington, Massachusetts 02173-0073, U. S. A.

Abstract Large-area, device-quality Si films on SiO$_2$ have been prepared by zone-melting recrystallization using graphite strip heaters. A composite SiO$_2$/Si$_3$N$_4$ encapsulation layer prevents agglomeration of the molten Si, insures a smooth film surface, and induces (100) texture. The recrystallized films contain widely-spaced grain boundaries, which can be eliminated by seeded growth techniques, and many sub-boundaries within each grain. Sub-boundaries can be entrained along parallel lines underneath a photolithographically defined optical absorber or reflector pattern. Extensive electrical measurements have been made on the recrystallized films. Sub-boundaries have no significant effect on MOSFET device performance, and high-yield CMOS test circuits have been made in films on 2-inch-diameter wafers. Radiation-hardened CMOS devices, lateral bipolar transistors, and dual-gate MOSFETs have been fabricated in recrystallized films.

1. Introduction

Considerable efforts have recently been directed toward the development of Si-on-insulator (SOI) technology for producing high-quality, device-worthy Si films on insulating substrates.[1-4] These efforts have been motivated by the potential of thin-film devices for achieving higher packing density, speed and radiation resistance than bulk devices, and also by the potential of SOI structures for accomplishing 3-dimensional integration of electronic circuits. Of the several SOI approaches currently under investigation, one of the most promising is zone-melting recrystallization, in which the grain size of a polycrystalline Si film on an insulating substrate is greatly increased by the passage of a narrow molten zone. In this chapter we discuss the graphite-strip-heater zone-melting-recrystallization technique being developed at Lincoln Laboratory and review the structural and electrical properties and device applications of zone-melting-recrystallized Si films.

2. Recrystallization Technique

Various methods have been used to produce the molten zone. Focused laser or electron beams have sufficient power density for melting a Si film, even with the substrate kept at relatively low temperature ($\sim 600°C$). On the other hand, if the substrate is heated to near the melting point of Si by some auxiliary means, a resistively heated wire or graphite strip[5-7] or a focused lamp [8,9] can be used to supply the additional heat necessary to achieve melting.

The configuration of the movable graphite-strip-heater oven we have used for zone-melting recrystallization is shown in Fig. 1. Samples have consisted of fine-grain polycrystalline Si films on an insulating substrate or layer (usually a SiO_2 layer on a single-crystal Si wafer) and an encapsulation layer over the poly-Si. The insert of Fig. 1 depicts a cross section through a typical sample, which consists of a layer of thermal SiO_2 0.1 to 1 μm thick on a Si wafer, a poly-Si film typically 0.5 μm thich deposited by low-pressure chemical-vapor deposition (LPCVD), a 2μm layer of CVD SiO_2, and a 30-nm layer of sputtered Si_3N_4. The sample is heated to 1100–1300°C by the lower strip heater, and additional radiative heating from the upper strip produces a narrow molten zone in the Si film directly below it. This molten zone is then scanned across the sample by moving the upper strip heater, typically at a speed of 1 mm/sec. The experiments are performed in a flowing Ar gas ambient. Several graphite-strip-heater

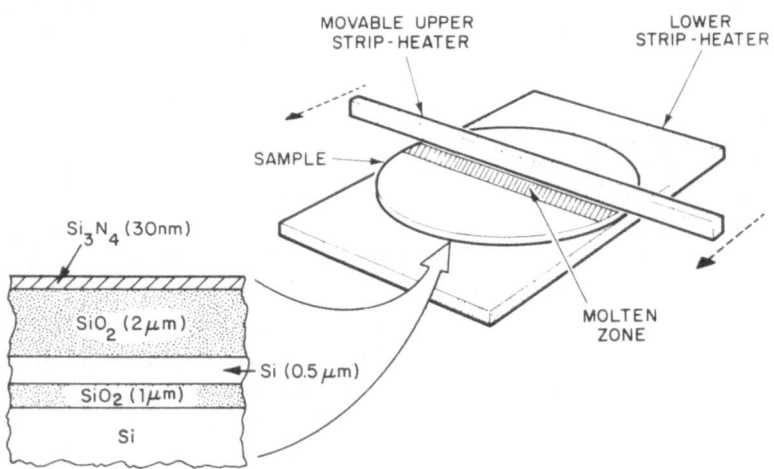

Fig. 1. Schematic diagram of sample and strip heater used in zone-melting recrystallization of Si films.

ovens have been constructed. The most recent one can routinely recrystallize flims up to 3 inches in diameter.[10]

3. Encapsulation

Molten Si does not wet SiO$_2$ well and has a tendency to agglomerate into droplets during recrystallization. Agglomeration can be reduced by keeping the molten zone narrow, 100 μm wide or less, which can be achieved in laser or electron beam recrystallization. In this case the zone can be confined by the surrounding solid Si, which is better wet by molten Si.[11] In strip-heater recrystallization, the molten zone is considerably wider, 1–2 mm, and an encapsulation layer is needed to stabilize the melt. We have found that a composite SiO$_2$/Si$_3$N$_4$ encapsulant, as illustrated in Fig. 1, is effective in preventing agglomeration. [6,7]

The thin Si$_3$N$_4$ layer on top of the SiO$_2$ is important. As shown in Fig. 2, without the Si$_3$N$_4$ the SiO$_2$ breaks up and the Si film forms many droplets, whereas smooth films are obtained with the SiO$_2$/Si$_3$N$_4$ cap. The role of Si$_3$N$_4$ in preventing agglomeration is not well understood. A thin Si$_3$N$_4$ layer (6 nm thick) deposited directly on top of the poly-Si enhances melt stability and improves the surface morphology of Si films during laser recrystallization.[12] We have found that reversing the sequence of Si$_3$N$_4$ and SiO$_2$ layers in the cap is also effective in preventing agglomeration during strip-heater recrystallization. These observations suggest that for samples with Si$_3$N$_4$ on top of the SiO$_2$ cap, N may be incorporated in the SiO$_2$ layer and at the SiO$_2$/Si interface during recrystallization, enhan-

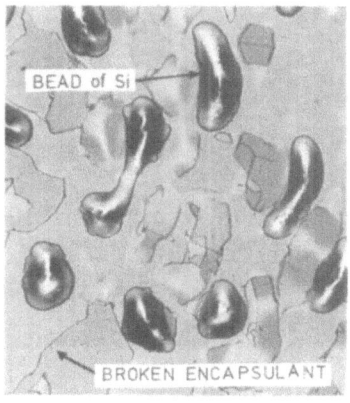

a   b

Fig. 2. Photomicrographs of zone-melting-recrystallized Si films encapsulated with (a) SiO$_2$ only (b) composite SiO$_2$/Si$_3$N$_4$ cap.

cing the wetting of SiO_2 by molten Si. However, N was not detected [13] either in the SiO_2 cap or at the SiO_2–Si interface by our Auger analysis (which has a detection sensitivity of ~0.1 atomic %). Another possible explanation is that the Si_3N_4 layer enhances the mechanical strength of the 2-μm-thick SiO_2 layer. The composite cap provides suitable stress which maintains continity of the Si films during recrystallization. Regardless of the mechanism, the cap structure shown in Fig. 1 produces most satisfactory results in terms of both good material properties and electrical characteristics.

In addition to preventing agglomeration, the encapsulation layer also serves the purpose of achieving a smooth surface and inducing (100) crystalline texture in the recrystallized films. The former aspect is illustrated in Fig. 3, which shows the optical micrographs of recrystallized Si films for two thicknesses of the SiO_2/Si_3N_4 encapsulant. For 0.2 μm SiO_2/30 nm Si_3N_4, the Si film is highly faceted, whereas it is smooth for the 2 μm SiO_2/30 nm Si_3N_4 encapsulant. [14] It is obvious that the thicker SiO_2 cap is mechanically stronger and therefore more effective in maintaining smooth surface morphology.

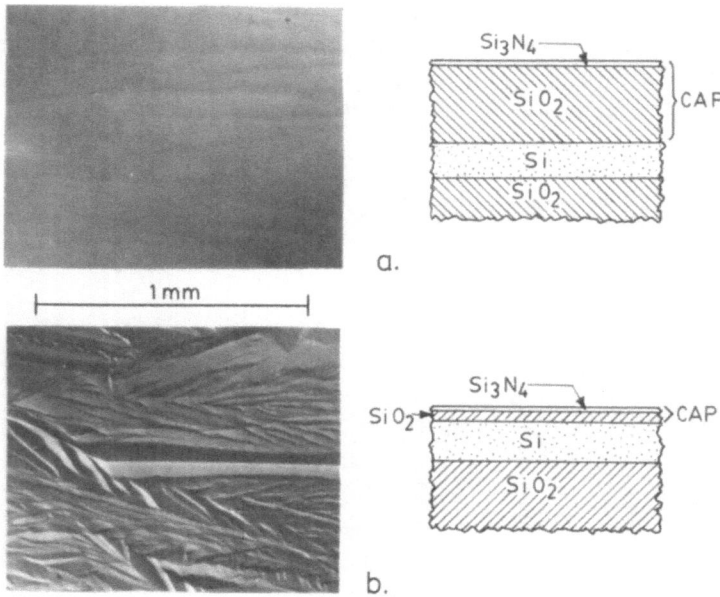

Fig. 3. Photomicrographs of recrystallized Si films with SiO_2/Si_3N_4 encapsulant in which the SiO_2 layer was (a) 2 μm thick, (b) 0.2 μm thick. In both cases the Si_3N_4 layer was 30 nm thick.

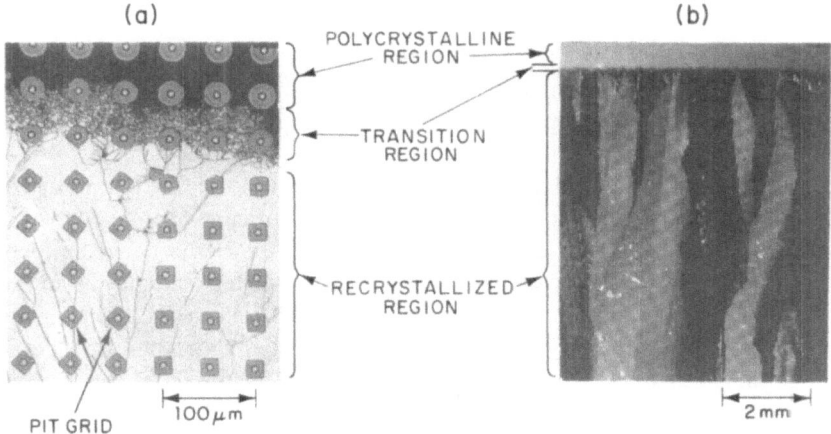

Fig. 4. Photomicrographs of the starting end of a recrystallized Si film. A grid array of etch pits has been etched into the film, which has also been etched to reveal the grain boundaries and sub-boundaries. Reflection from the etch pit facets makes crystals with different orientations stand out distinctly. (a) High magnification, (b) low magnification.

4. Texture and Orientation

The zone-melting-recrystallized Si films contain highly oriented crystals. Figure 4(a) is a high-magnification optical micrograph showing the starting end of a recrystallized film that has an etch pit grid [7,15] and has also been etched to delineate grain boundaries and other defects. Three areas are discernible: an area of very fine grained Si that was not melted, a trasition region with larger grains, and an area where the Si film was completely melted and recrystallized. The square shape of the pits indicates (100) texture, and the diagonals of the pits are parallel to the ⟨100⟩ direction. Away from the transition region there are two large grains, both with (100) texture, that are rotated relative to one another by about 45° in the plane of the substrate. Figure 4(b) is a micrograph showing a larger area of the same recrystallized sample. The variation in the intensity of light reflected from the facets of the etch pits makes the individual grains stand out distinctly. The grains have a width of 0.5–2 mm, and their length can be increased indefinitely as long as growth is not interrupted. The figure shows that the recrystallization is seeded from the transition region. Within a few millimeters of this region, grains that have (100) texture and their ⟨100⟩ axes close to the scanning direction of the upper strip heater dominate the film.

The well-oriented growth of zone-melting-recrystallized films depends on the formation of the (100) texture, which is believed to be induced by the SiO_2 cap. The (100) texture is observed both for material solidified from the molten zone and for the adjacent transition region [see Fig. 4(a)]. The transition region, which is typically between 0.1 mm and a few millimeters wide, is formed between the randomly-oriented fine-grained poly-Si (grain size (0.5 μm) and the initial molten zone. Seeds for solidification of the molten zone are provided by the transition region.

The manner in which the cap induces (100) texture is not well understood. It has been suggested [14] that the significant change in texture that occurs in the transition region is associated with the coexistence of solid and liquid due to partial melting. The unmelted crystallites in the transition region may be predominantly of (100) texture and then grow during the course of recrystallization. The coexistence of solid and liquid phases in Si heated by laser radiation has been observed by several investigators.[16,17] Other explanations include heterogenous nucleation of (100) crystals from the melt onto SiO_2[11] and stress-induced texturing.[13]

The explanations for (100) texture are all based on the assumption that the interfacial energy between Si and SiO_2 is lowest for the (100) planes, making growth of (100) crystals energetically favorable. In an attempt to test this hypothesis, we have conducted experiments in which a 0.5-μm-thick poly-Si film was zone melted over a thermal SiO_2 layer that had parallel slit openings etched down to a (111)-oriented Si substrate. The distance between parallel openings was 250 μm, and the encapsulation layer over the Si film was 2 μm SiO_2/30 nm Si_3N_4. When the heater was scanned in a direction perpendicular to the slit openings, (111) Si films were obtained. However, when the heater was scanned parallel to the openings, as illustrated in Fig. 5, the growth of (111) crystallites was quickly terminated as a result of fast lateral growth of adjacent (100) crystallites that originated in transition regions located between adjacent slit openings.

5. Grain Boundaries and Sub-Boundaries

In films recrystallized by zone melting we can distinguish two types of crystallographic boundaries: grain boundaries and subgrain boundaries (or sub-boundaries). The range of grain-boundary angles is reduced with distance from the transition region, but is not eliminated. Single-grain films can be obtained by seeding from a single-crystal substrate [5] or using an external seed [18] as in the LESS technique, by cross seeding, [7], or by zone melting of patterned films.[19] However, sub-boundaries remain in all these films.

Sub-boundaries, which consist primarily of arrays of dislocations, [20] have crystallographic angular deviations of one degree or less. They originate at the interior corners of the faceted growth fronts, as illustrated

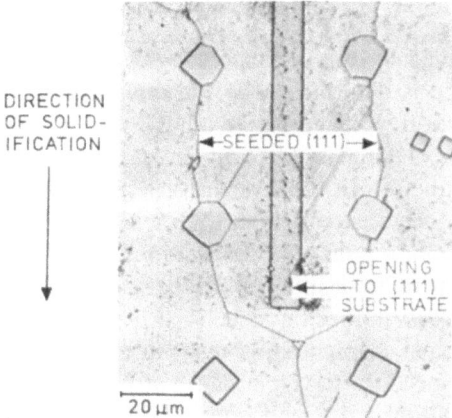

Fig. 5. Photomicrograph of recrystallized Si film, showing that growth from a (111) seed is rapidly eliminated by growth from (100) seeds. Etch pits indicate crystal texture and orientation. Note the approximate 6-fold symmetry of the portions of pits located in the region recrystallized from the (111) seed.

by the photomicrograph shown in Fig. 6. To obtain this micrograph, the molten zone was rapidly quenched with a jet of He gas in order to reveal the solid-liquid interface at the trailing edge of the zone. Etching studies

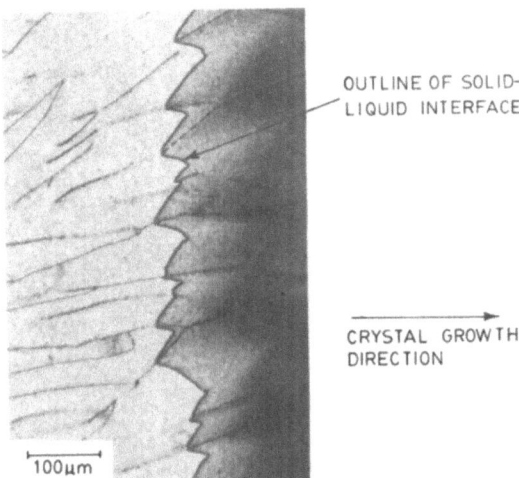

Fig. 6. Photomicrograph of recrystallized Si film that was quenched by a He gas jet during zone melting.

have shown that the growth front facets are close to (111) planes. Since the angle between these planes is fixed, the separation between sub-boundaries is determined by the distance to which the facets extend into the melt. This distance is determined by the temperature gradient in the heater scanning direction.

The temperature gradient depends on the scanning speed. Increasing this speed causes a reduction in the temperature gradient, which increases the distance that the facets extend into the melt and therefore increases the sub-boundary spacing. This effect is demonstrated in Figs. 7(a) and 7(b), which are optical micrographs of the pattern of sub-boundaries in 0.5-μm thick Si films recrystallized at speeds of 0.2 and 2 mm/sec, respectively. The sub-boundary spacing cannot be indefinitely increased by increasing the scanning speed, however, because at very high scanning speeds control of recrystallization is difficult and dendritic growth often occurs. Most of our experiemtns have been performed at a speed of 1 mm/sec.

The temperature gradient at the growth interface also depends on the heat dissipation, which is a function of both the gas ambient and the sample structure. Recrystallization in He gas, which has higher thermal conductivity than Ar, results in an increased temperature gradient and therefore a closer sub-boundary spacing. Increasing the thickness of the Si film increases the total heat of fusion that has to be dissipated, and increasing the thickness of the underlying SiO_2 layer reduces the rate of heat dissipation to the substrate; both result in a decreased temperature gradient and hence an increased sub-boundary spacing.

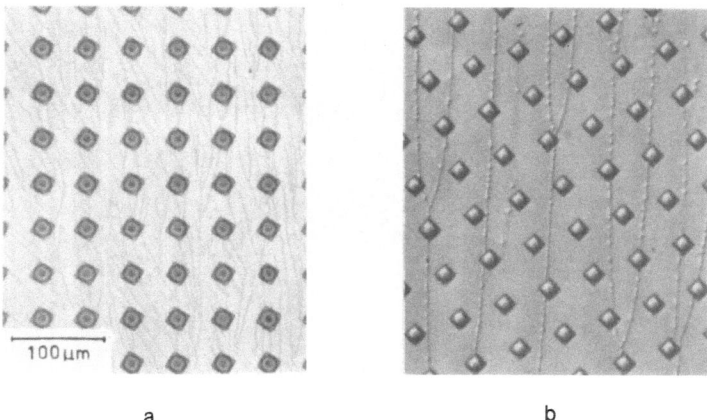

a   b

Fig. 7. Photomicrographs showing sub-boundaries in 0.5-μm-thick Si films recrystallized at speeds of (a) 0.2 mm/sec and (b) 2 mm/sec.

6. Sub-Boundary Entrainment

So far we have not been able to eliminate sub-boundaries over large areas. Several methods have been proposed to reduce sub-boundary density, such as repeated melting and resolidification [21] and zone melting films that have been patterned into stripes or islands. [11] Both techniques have been used in laser recrystallization to produce defect-free Si regions 100 μm in size.[11] We have investigated[22] a technique for entraining sub-boundaries by the use of photolithographically defined patterns that modulate the solidification front. The localization of sub-boundaries at well-defined positions gives one the option of building electronic devices in the defect-free regions between the sub-boundaries. A similar technique employing patterned antireflection coatings in laser crystallization has been used [23] to achieve grain boundary localization.

Since the sub-boundaries originate at the interior corners of the faceted growth front, sub-boundaries can be entrained by controlling the positions of the facet intersections. The technique we have used for this purpose is to spatially modulate the temperature at the solid-liquid interface by forming a grating above the Si film that locally enhances either the absorption or reflection of the radiation incident from the upper heater. Figure 8(a) is a schematic cross-section diagram illustrating the use of a patterned optical absorber for sub-boundary entrainment. Since the Si is

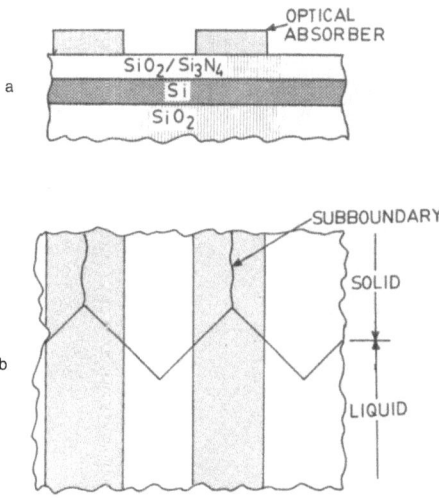

Fig. 8.   Schematic diagrams of (a) cross section of sample with grating of optical absorber stripes on top of encapsulation layer (b) entrainment of sub-boundaries under the middle of the optical absorber stripes.

hottest under the middle of each stripe, this is the last place to solidity. For this reason, the interior corners of the growth front, and hence the sub-boundaries, align with the middle of the absorber stripes, as depicted in Fig. 8(b). Some results obtained by using carbonized photoresist as an optical absorber are shown in Fig. 9(a). The Si film was 1 μm thick, the substrate was 1 μm of SiO_2 over a (100) Si wafer, the encapsulant was 2 μm of SiO_2 and 30 nm of Si_3N_4, and the absorber pattern was a grating of stripes 50 μm wide with a 100 μm period. The scanning speed was 0.5–1 mm/sec. The Si between the entrained sub-boundaries is free of dislocations, as determined by chemical etching, and no carbon contamination was detected by Auger analysis, which has a detection limit of about 10^{19} atom/cm^3. Figure 9(b) shows a typical pattern of sub-boundaries from a control sample that did not have an absorber pattern.

Figure 10 illustrates the use of a grating of thin Si stripes imbedded in the encapsulation layer to achieve entrainment. Because of the high reflectivity of Si at the temperatures used in zone melting, in this case the sub-boundaries are entrained between the stripes.

7. Topographic Imperfections

Zone-melting-recrystallized Si films exhibit several types of topographic inperfections. One type is wafer warpage, which must be minimized because the characteristics of the photolithographic process used in VLSI fabrication impose stringent requirements on wafer flatness. We have measured [24] wafer flatness with a laser interferometer, which gives an interference

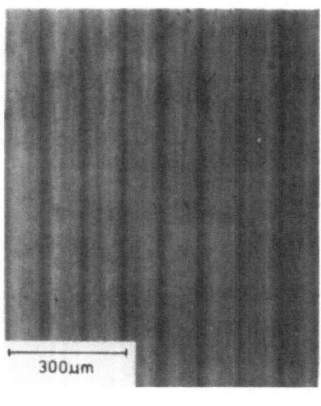

a   b

Fig. 9.   (a) Sub-boundaries in Si film recrystallized with grating of carbonized photoresist stripes on top of encapsulation layer. (b) Sub-boundaries in control film recrystallized without absorber grating.

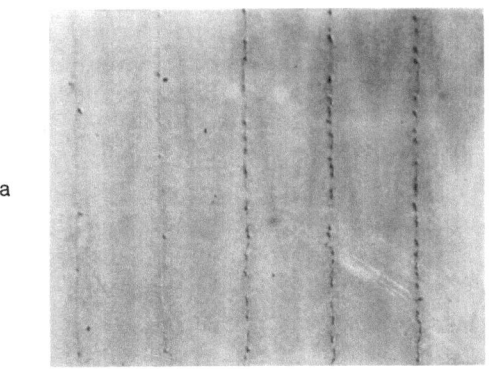

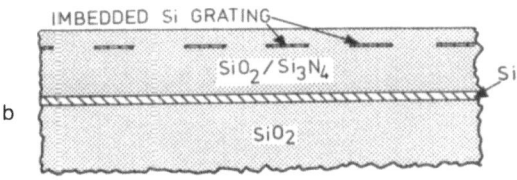

Fig. 10 (a) Sub-boundaries in Si film recrystallized with grating of Si reflector stripes imbedded in encapsulation layer. (b) Schematic cross-section diagram of sample with imbedded Si reflector stripes.

pattern corresponding to a topographic map of the sample surface. Before recrystallization the overall warp, peak to valley, of a 2-inch-diameter sample placed on a vacuum chuck is typically 2–4 μm, a value comparable to that for a bulk Si wafer after device processing. After recrystallization the samples remain relatively flat over most of their surface, but occasionally there are one or two high points giving rise to a total warp of over 25 μm. The high points result from sample imperfections that protrude from the back side of the Si substrate and prevent the wafer from conforming to the flat surface of the vacuum chuck. When these imperfections are removed by mechanically lapping the back side of the sample, the overall warp decreases dramatically, to less than 6 μm. To simulate the effects of high-temperature device processing, one recrystallized sample was annealed for one hour at 1000°C and for another hour at 1100°C. The overall flatness was actually slightly improved. Similar results have been achieved for 3-inch-diameter wafers with total warp less than 8 to 12 μm.

Because of the thermal stress inherent in the recrystallization process, the Si wafers often show slip planes, although the recrystallized Si films themselves do not. While slip planes do not interfere with our device processing procedures, they can increase sample breakage during handling, and careful thermal cycling must be used to minimize their formation.

Another type of topographic imperfection is the surface protrusion. Small pools of molten Si are often trapped by the advancing growth front. When these molten pools solidify, the Si expands to form protrusions, as shown in Fig. 11. The protrusions generally lie along sub-boundaries, and their density increases with increasing sub-boundary spacing. By increasing the temperature gradient, which decreases the sub-boundary spacing, protrusions can be greatly reduced. Thus samples with a thinner SiO₂ layer underneath the Si film generally have lower protrusion densities. Two-inch-diameter films nearly free of protrusions have been obtained.

8. MOS Devices and Circuits

The electrical characteristics of zone-melting-recrystallized Si films have

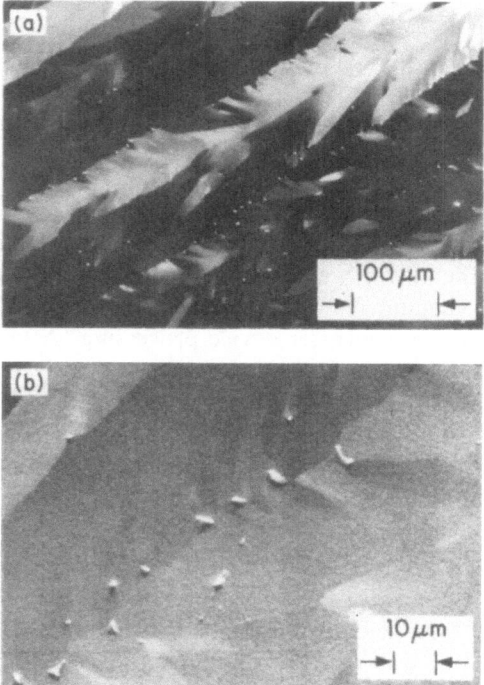

Fig. 11   (a) Micrograph of recrystallized Si film taken with a scanning electron microscope operated in the backscattering mode. The Si₃N₄/SiO₂ cap has been removed. (b) Magnified view of a small area of the film showing protrusions.

been extensively studied.[25 – 30] Thin-film resistors and n-channel MOSFETs have been used [27] to evaluate the majority carrier transport properties. The sub-boundaries, which are the predominant defects in the material, have very little effect on majority carrier transport and MOS device performance. N-channel MOSFETs with electron transport either parallel or perpendicular to the sub-boundaries show nearly identical surface electron mobility. This is consistent with the measurements on thin-film resistors, which indicate that the sub-boundary trapping state density is low enough that sub-boundaries do not produce a large potential barrier to impede carrier transport. The grain boundaries in the films, however, have been found [31] to cause a significant reduction in conductivity and degradation of MOSFET performance.

Although sub-boundaries do not have a significant effect on majority-carrier transport, MOSFET performance might be degraded by enhanced dopant diffusion occurring along the sub-boundaries during device fabrication. For narrow-channel MOSFETs fabricated in laser-recrystallized Si films, dopant diffusion along grain boundaries has been found to result in shorting between source and drain or in high source-to-drain leakage currents.[32] We have fabricated n-channel MOSFETs with gate lengths ranging from 1.5 to 45 μm in zone-melting-recrystallized films. Source and drain dopant activation was accomplished by annealing at 900°C for 30 min. The leakage current was in the picoampere range and nearly independent of gate length. This result indicates that dopant diffusion is not greatly enhanced by the sub-boundaries.

The most important application of SOI materials is in high-density and high-speed integrated circuits. Higher packing density can be achieved because of the simplicity of device isolation, and higher operating speeds result from the reduction in parasitic capacitance. As we have previously reported, [25,28] both n- and p-channel MOSFETs fabricated in the recrystallized Si films exhibit electrical characteristics comparable to those of single-crystal Si devices. In order to carry out a more critical evaluation of zone-melting-recrystallized SOI films for IC applications, we have designed [30] a CMOS test circuit chip for fabrication in these films. The test chip, which is based on a 5 μm design rule, contains n- and p-channel transistor arrays, ring oscillators, inverter chains, and various test devices for process control. The objectives of utilizing this design are to assess the uniformity of the SOI films and to determine the speed of SOI/CMOS circuits. Film uniformity is required for obtaining a high yield of functional circuits, while speed is a key parameter in deciding between alternative material technologies for VLSI. The CMOS test chips were fabricated on SOI structures consisting of a 0.5-μm-thick recrystallized Si film, a 1-μm-thick SiO$_2$ layer, and a 1 Ω-cm p-type Si$\langle 100 \rangle$ wafer 2 inches in diameter. The fabrication process used involves a total of six photomask steps with poly-Si gate and self-aligned ion-implanted source and drain. Figure 12 is a photomicrograph of a finished SOI/CMOS chip, which measures about 3×4 mm. About 100 such chips were fabricated on each of three wafers. A 31-stage ring oscillator with fan in and fan out of one and a 231-stage inverter chain are located at the upper left. A similar circuit, rotated by 90°, is placed at the upper right. The purpose of the rotation is to examine the effects of sub-boundaries on circuit performance. Carrier transport is approximately parallel to these boundaries in one circuit and approximately perpendicular to them in the other. Test transistors, single-stage inverters, gated diodes, capacitors and test patterns are located in the middle portion of the chip. Two n-channel transistor arrays consisting of 360 or 533 parallel devices and a p-channel array consisting of 460 parallel transistors are located at the lower left. The individual transistors, which have a 5 μm gate length and 20 μm gate width, are

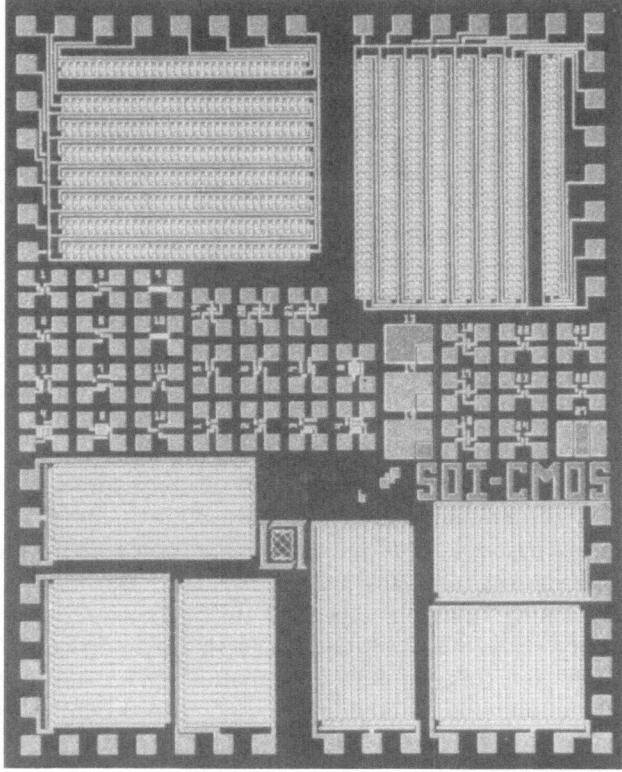

Fig. 12.   Photomicrograph of SOI/CMOS test chip fabricated in zone-melting-recrystallized Si film.

spaced 5 μm apart. Three similar arrays, rotated by 90°, are located at the lower right.

To evaluate the uniformity of the SOI films, we have investigated the performance of all the transistor arrays on a wafer with 98 chips. Since the transistors in each array are connected in parallel, failure of a single device results in failure of the entire array. Of the 588 arrays, 490 were functional while 98 failed because of source-to-drain short or open circuits. For 62 of the inoperable devices, localized metallization defects such as incomplete etching of the Al or poor contacts were found by microscopic inspection. Thus the overall yield of functional arrays exceeds 90% when the obvious fabrication defects are discounted. The total number of transistors in all the arrays is 2.65×10^5. If it is assumed that each of the failed arrays contains one defective device, the transistor failure rate is 3.7×10^{-4}, or only 1.4×10^{-4} if the known fabrication defects are taken into account.

The operating characteristics of the functional transistor arrays are quite uniform from chip to chip. For the three types of arrays in which carrier transport is approximately parallel to the sub-boundaries, the average values and standard deviations for the transconductance are as follows: p-channel (measured at $V_D = V_G = -5$ V), 52 ± 2 mS; 360-transistor and 533-transistor n-channel (measured at $V_D = V_G = +5$ V), 78 ± 2 and 115 ± 3 mS, respectively. The arrays in which carrier transport is perpendicular to the boundaries have similar characteristics, except that the transconductance values are 5% lower.

Measurements have also been made on all of the 31-stage ring oscillators on the 98-chip wafer. For the two orientations differing by 90°, 82 of the 98 oscillators in each set are functional. Again, most of the failures can be attributed to obvious metallization defects. The output waveform and operating characteristics of a typical functional oscillator are shown in Figs. 13(a) and (b), respectively. The circuit starts to oscillate at a supply voltage V_D of 1.5 V. at $V_D = 5$ V, the swiching delay time and dissipated power are respectively 2 ns and 0.13 mW per stage, for a power-delay product of 0.26 pJ. The operating speed can be attributed to the high carrier mobilities in the recrystallized Si films and the reduced parasitic capacitance of the SOI structure. Again the two sets of oscillators are similar, indicating that the sub-boundaries, despite their large number, do not have a significant effect on circuit performance.

The yield of inverter chains on the test chips is comparable to the yield of ring oscillators. The performance of the functional chains has been tested by two methods: (1) an input signal is supplied, and the propagated signal is monitored at different output stages, and (2) the input and output are connected together so that the circuit functions as a ring oscillator. The chains exhibit normal inverter characteristics and switching

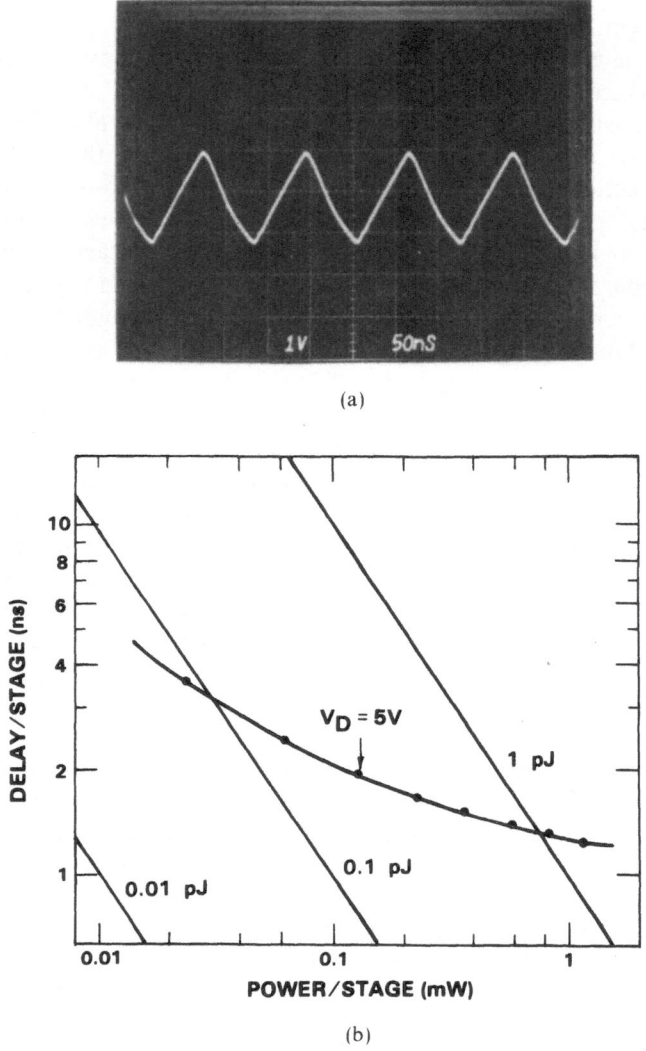

(a)

(b)

Fig. 13. (a) Ouput waveform and (b) switching delay time of 31-stage SOI/CMOS ring oscillator as a function of power dissipated per stage.

delay times similar to those measured for the ring oscillator circuits.

For each of the other 2-inch-diameter SOI wafers on which CMOS chips were fabricated, we have tested the transistor arrays, ring oscillators, and inverter chains on about 10 chips distributed over the area of the wafer. All the functional circuits measured are comparable in performance to those on the wafer for which all 98 chips were tested.

9. Radiation Hardness

SOI/MOS devices have several advantages over their bulk Si counterparts as components for radiation-hardened circuits: reduced transient photocurrent, elimination of latch-up, increased speed and packing density. However, SOI devices can be degraded by radiation-induced leakage current, which is caused by back-channel conduction due to positive charge trapping at the interface between the Si films and the SiO$_2$ layer underneath. For n-channel SOI/MOSFETs, it has been demonstrated[33,34] that the effects of charge trapping in the lower SiO$_2$ layer can be largely suppressed by applying a negative bias to the Si substrate during irradiation. In experiments with a bias voltage of -15 V, subthreshold leakage currents of less than 0.2 pA/μm (channel width) were obtained [33] for n-channel SOI/MOSFETs subjected to ionizing doses up to 10^6 rad (Si).

Applying a negative bias to the Si substrate during irradiation of SOI/MOSFETs can greatly reduce the effect of ionizing radiation on the leakage current of n-channel devices because such a bias can reduce the quantity of radiation-induced positive charge trapped at the lower Si-SiO$_2$ interface and also counteract the tendency of this charge to induce an n-type inversion layer in the Si film. For p-channel SOI/MOSFETs, however, a negative substrate bias can have two adverse effects on device characteristics in the absence of radiation. First, a depletion region will be formed at the back channel, which can result in space-charge interaction with the front channel and cause a shift in threshold voltage. Second, for sufficiently large negative substrate bias, a weak inversion layer will be formed at the lower Si-SiO$_2$ interface that can lead to an increase in subthreshold leakage current. Furthermore, irradiation actually tends to improve the performance of p-channel devices because their leakage current is decreased by positive chage trapped at the lower Si-SiO$_2$ interface; applying a negative substrate bias can reduce this beneficial effect. Consequently, to utilize negative substrate biasing for protecting SOI/CMOS devices against ionizing radiation we must determine an optimum bias voltage for the combined performance of the n- and p-channel devices.

CMOS inverters fabricated in the zone-melting-recrystallized Si films were irradiated with a beam of 1.5-MeV electrons from a Van de Graaff generator. Figure 14 is a schematic diagram of the device structure. The electron doses ranged from 3×10^{10} to 3×10^{13} cm^{-2}, corresponding to ionizing doses from 10^3 to 10^6 rad(Si). The bias voltages during irradiation were $V_{DD} = 5$ V, $V_{IN} = 0$ or 5 V, and $V_B = 0$, -5, -7.5, or -10 V, where V_{DD} is the supply voltage and V_{IN} is the input voltage. Subthreshold leakage current and threshold voltage measurements were performed within 20 min after exposure. In all cases the V_B value was the same during

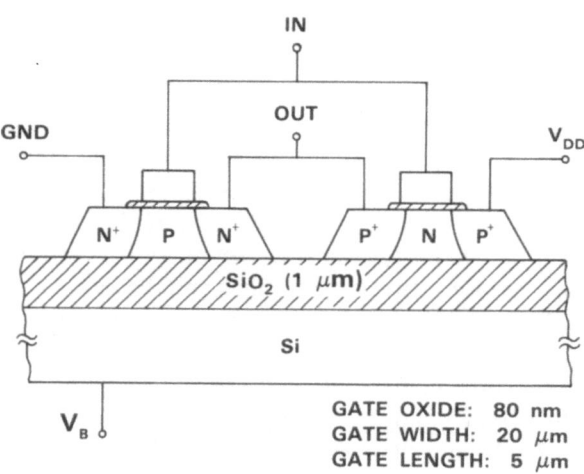

Fig. 14. Schematic structure of SOI/CMOS inverter fabricated in zone-melting-recrystallized Si film on SiO₂-coated Si substrate.

the measurements as during irradiation.

For inverters with $V_B = 0$, large n-channel leakage currents ($> 10^{-8}$ A) were observed after irradiation to doses of 10^5 rad(Si) or above, while there was no increase in p-channel leakage current. On the other hand, for $V_B = -7.5$ or -10 V, the n-channel leakage current remained low after irradiation but the p-channel leakage current was 2–3 orders of magnitude higher than for $V_B = 0$ before and after irradiation. Best results were obtained with $V_B = -5$ V. Figure 15 shows the subthreshold characteristics of n- and p-channel devices for $V_{IN} = 0$ V and $V_B = -5$ V during irradiation. These characteristics were measured separately after irradiation for the n- and p-channel devices, with the same V_B and with drain voltages of 5 and -5 V, respectively. Leakage currents of less than 0.1 pA/μm (channel width) were observed for both devices up to 10^6 rad(Si) dose. For $V_{IN} = 5$ V and $V_B = -5$ V during irradiation, the results are shown in Fig. 16. Both n- and p-channel leakage currents remained low after 10^5 rad(Si); after 10^6 rad(Si) there was a large increase in n-channel leakage current but hardly any change in p-channel leakage current. The increase in n-channel leakage current is mainly caused by the threshold-voltage shift, which is known to occur most significantly for n-channel devices that are irradiated under positive gate bias.[35] This threshold shift should be markedly reduced by using radiation-hardened gate oxide and by fabrication procedures that produce less gate oxide degradation. Receutly, we have fabricated[36] SOI/CMOS devices with thin gate oxide of 10 nm thick and they exhibit excellent radiation resistance up to 10^7 rad (Si) with respect to leakage current, threshold voltage and transconductance.

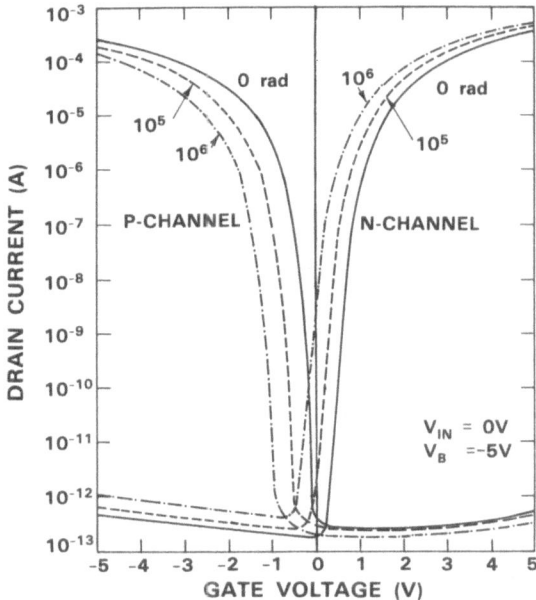

Fig. 15. Subthreshold characteristics of n- and p-channel devices with $V_{IN} = 0$ V and $V_B = -5$ V during irradiation. These characteristics were measured separately for the n- and p-channel devices, with the same V_B and with drain voltages of 5 and -5 V, respectively.

We therefore believe that a radiation-hardened SOI/CMOS technology can be developed for achieving high packing density and high-speed operation.

10. Bipolar Devices

Another possible of application for SOI material is in the area of bipolar devices, which have the potential for high speed operation. A study[29] using the pulsed MOS capacitance technique indicated that the minority carrier lifetimes in the zone-melting-recrystallized films are close to the microsecond range, suggesting the possibility of employing these films for bipolar devices. By a minor modification of the MOS processing procedure used for SOI films, we have fabricated a novel four-terminal device that can be operated either as a lateral npn bipolar transistor or as a conventional MOSFET. The individual devices are fully isolated, making it possible to achieve complete integration of bipolar and MOS devices with minimum interaction between them. This capability could be useful for many digital and analog integrated circuit applications.[37]

The SOI structures used for device fabrication consist of a 0.5-μm-thick recrystallized Si film on a 1-μm-thick layer of thermal SiO$_2$ on a

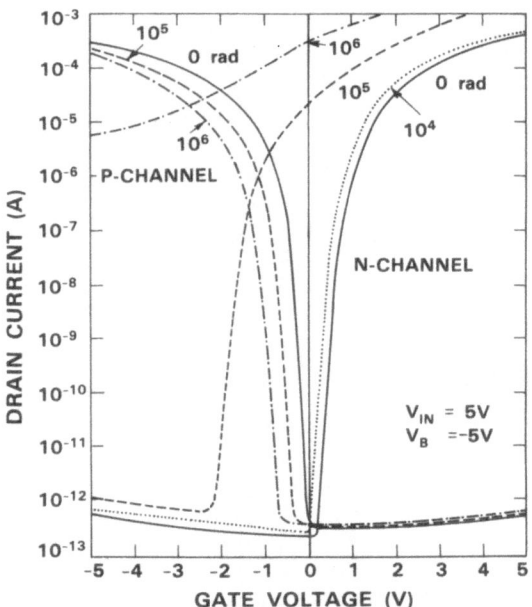

Fig. 16.   Subthreshold characteristics of n- and p-channel devices with $V_{IN} = 5$ V and $V_B = -5$ V during irradiation.

single-crystal Si substrate. Figure 17(a) is a schematic cross section of the device structure. The active area was first defined by local-oxidation-of-Si (LOCOS) isolation and doped with B by ion implantation. Gate oxide 100 nm thick was grown and poly-Si film was then deposited and defined to form the gate. A low-dose As^+ implant was performed for collector doping. The emitter and collector contact regions were doped by a high-dose As^+ implant, with photoresist used as a mask to protect the collector area. The contact window to the base was then opened and doped with B by ion implantation to reduce contact resistance. After CVD SiO_2 passivation and high-temperature drive-in, contacts to the emitter (source), base, collector (drain), and gate were opened and metallized with Al. The device was completed by using H_2 sintering to reduce the density of SiO_2–Si interface states. Figure 17(b) is a photomicrograph of a finished device, which has a nominal gate width (emitter length) of 50 μm and gate length (base width) of 3 μm.

When operated as an n-channel MOSFET, the device exhibits well-behaved enhancement-mode characteristics with a threshold voltage of $+1.3$ V and surface electron mobility of ~ 600 cm^2/V-s. Figure 18(a) shows the source-drain (emitter-collector) I-V characteristics obtained for a device with base and gate contacts floating. The drain (collector) junction ex-

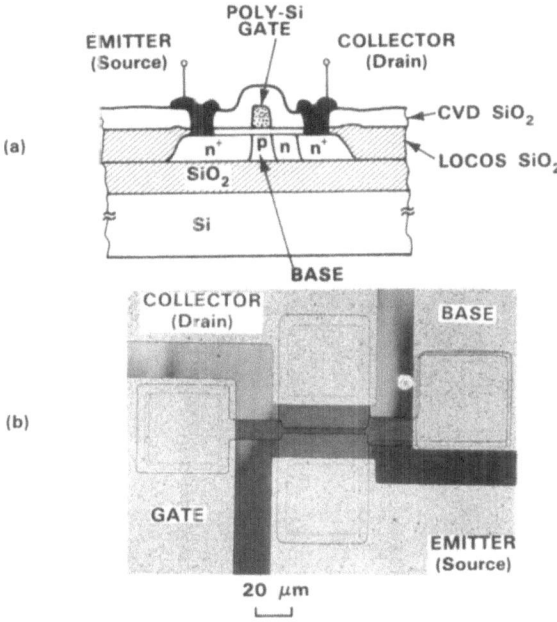

Fig. 17. (a) Schematic cross section of lateral bipolar/MOS transistor fabricated in zone-melting-recrystallized Si film. (b) Photomicrograph (top view) of a finished device.

hibits sharp breakdown, with the breakdown voltage V_B exceeding 15 V, compared to 10–15 V for the SOI/MOSFETs we previously fabricated in zone-melting-recrystallized films. The increase in drain V_B is presumably due to a reduction in impact ionization, which is expected because the decreased drain doping concentration ($\sim 1 \times 10^{17}$ cm^{-3}) in the new device results in lowering and spreading of the electrical field in the drain pinch-off region.[38] As shown in Fig. 18(a), the source (emitter) junction of the new device exhibits soft breakdown, with V_B exceeding 10 V. The asymmetry in breakdown properties between the emitter and collector junctions is due to the difference in their doping profiles. The emitter-base junction has well-behaved forward I-V characteristics (not shown), which follow the usual linear-log relation with an ideality factor of ~ 1.2.

Figure 18(b) shows the transistor characteristics obtained for lateral bipolar operation. The common-emitter current gain is 18 at low collector current but falls off rapidly at high current because of current crowding. The gain is relatively high compared to that of conventional bulk bipolar transistors, [39] probably because the complete isolation in the SOI structure eliminates the vertical emitter current that is present in bulk devices, where it is injected into the substrate without contributing to transistor

(a)

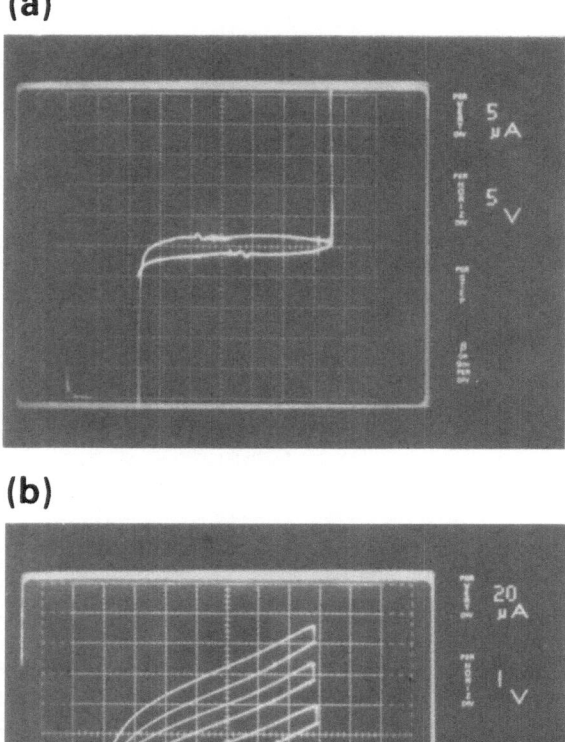

(b)

Fig. 18. (a) Source-drain (emitter-collector) *I-V* characteristics of device operated with base and gate contacts floating. (b) Common-emitter characteristics of device operated as lateral npn bipolar transistor.

action. The high current gain of the SOI device indicates a good minority-carrier diffusion length in the zone-melting-recrystallized film. Since the ratio of emitter to base doping is ∼ 10^3, it can be assumed that the emitter efficiency is about 1.[40] Taking the base width to be 3 μm, for the measured current gain of 18 the diffusion length calculated [40] for electrons in the base is 10 μm. This should be considered as a lower limit, since the current gain is reduced by surface recombination at the top and bottom SiO$_2$–Si film interfaces. An increase in current gain should be achieved by reducing the base width and optimizing the base doping.

The fully isolated SOI bipolar transistor, unlike its bulk counterpart, has no parasitic emitter-base and collector-base diodes. Furthermore, the vertical junction arrangement makes it possible to reduce the junction area and therefore the collector depletion capacitance. (For the present device this capacitance is 0.5 $fF/\mu m$ emitter length.) Consequently, the SOI bipolar transistor should be suitable for high-frequency operation. Its major disadvantage is the high base resistance resulting from the geometric configuration (see Fig. 17). Because of current crowding [40] this resistance will limit the current capability, although the problem should be less severe for smaller devices.

11. Dual-Gate MOSFETs

The usefulness of zone-melting-recrystallized SOI structures in the fabrication of three-dimensional (3-D) intergrated devices is limited by the high temperature required for the recrystallization process. However, some kinds of 3-D devices can be fabricated by using the recrystallization technique. To demonstrate this capability, we have designed and fabricated one such device, a dual-gate MOSFET. The fabrication process is shown schematically in Fig. 19. A CVD poly-Si film was deposited on a SiO_2-coated Si substrate and patterned by the LOCOS process, which also defined a buried poly-Si gate. After growth of gate oxide on the gate, a second CVD film was deposited and encapsulated with the standard SiO_2/Si_3N_4 cap (not shown in the diagram). The sample was processed in the strip-heater system to recrystallize both the second CVD film and the buried gate. A conventional poly-Si gate process was then used to fabricate MOSFETs in the top recrystallized Si film. Figure 20 shows a cross-section SEM micrograph of a finished n-channel dual-gate MOSFET. The gate oxides for the top and buried gates are 80 and 200 nm thick, respectively.

Since the two gates of the dual-gate MOSFET can independently modulate the channel, this device can be used as a two-input pass or NOR gate. Alternatively, if the two gates are connected together the device should have a high transconductance because both the top and bottom inversion channels contribute carrier conduction. In addition, the device should have a low leakage current because back-channel conduction can be eliminated by using the buried gate to control the bottom channel. Typical I-V characteristics for top-gate, buried-gate and dual-gate (both gates connected together) operation are shown in Fig. 21. The transconductance for dual-gate operation is equal to the sum of the tranconductances for top- and buried-gate operation.

DUAL-GATE MOSFETs

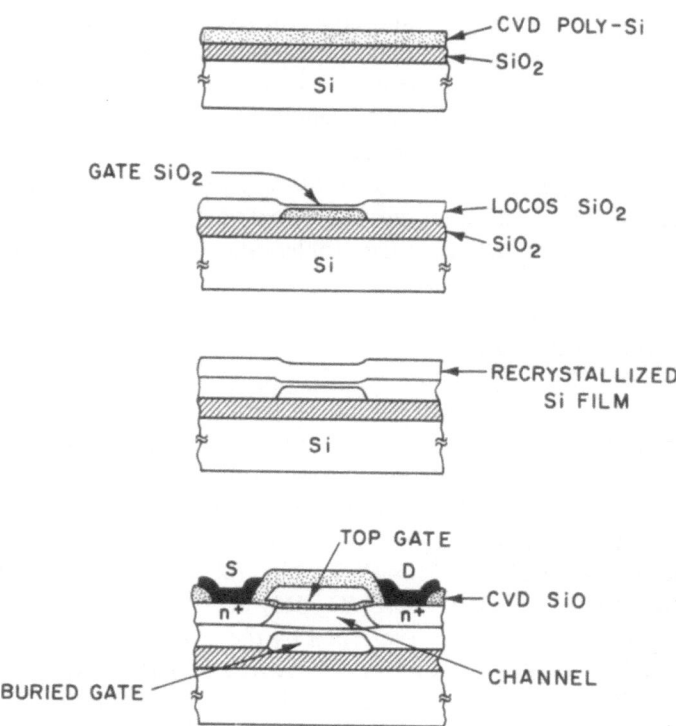

Fig. 19.   Schematic diagram showing fabrication process for dual-gate MOSFET.

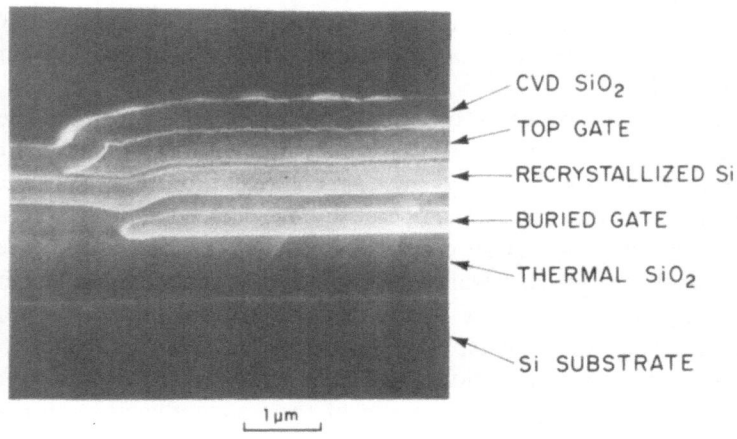

Fig. 20.   Cross-section SEM micrograph of finished dual-gate MOSFET.

TOP GATE

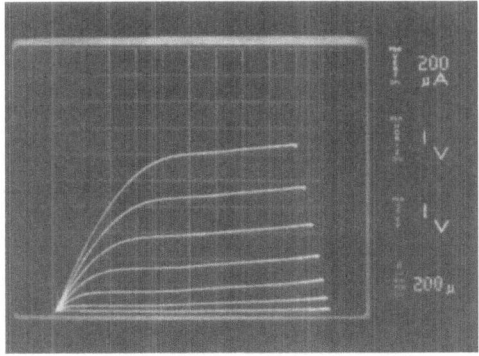

BOTTOM GATE

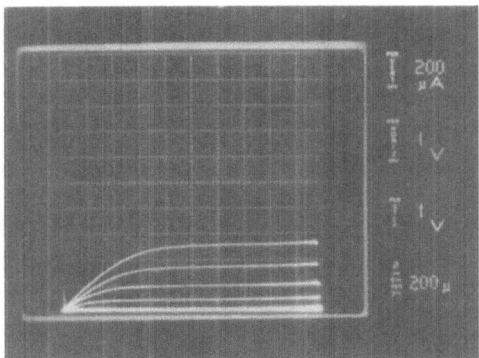

DUAL GATE

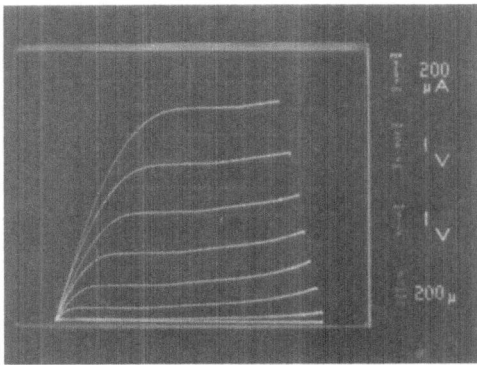

Fig. 21. Drain *I-V* characteristics of dual-gate MOSFET for (a) top-gate, (b) buried-gate, and (c) dual-gate (both gates connected together) operation.

12. Conclusion

A zone-melting recrystallization technique using graphite strip heaters has been developed for preparing SOI materials. Recrystallized Si films on 3-in.-diameter SiO_2-coated Si wafers can be routinely produced. Since systems employing graphite heaters can easily be scaled up, we anticipate that even larger wafers can be processed. A composite Si_3N_4/SiO_2 encapsulating layer is effective in preventing Si from agglomeration, thus ensuring smooth recrystallized surfaces. In addition, the cap induces a strong $\langle 100 \rangle$ texture. The recrystallized films are found to contain widely-spaced grain boundaries and many sub-boundaries within each grain. The grain boundaries can be eliminated by various seeding techniques.

We have performed extensive electrical measurements on our recrystallized films. Studies of thin-film resistors and n-channel MOSFETs have shown that the majority-carrier transport properties of zone-melting-recrystallized Si films are comparable to those of single-crystal Si. The sub-boundaries, which are the predominant defects in these films, have no significant effect on carrier transport or MOSFET characteristics. High yields of good-quality transistor arrays and ring oscillators have been obtained for CMOS test circuit chips. On the basis of initial experimental results on CMOS devices, the recrystallized films show great promise for radiation-hardened integrated circuits. Lateral bipolar transistors with current gain close to 20 have been fabricated in the recrystallized films. The zone-melting recrystallization technique has also been used for the fabrication of dual-gate MOSFETs with a 3-dimensional structure.

The results reported here show that zone-melting recrystallization yields SOI structures of sufficient quality to be seriously considered for utilization in VLSI circuits. It remains to be demonstrated that wafers suitable for conventional IC processing can be prepared by this technique. The characteristics required for employing large-area wafers, such as film uniformity and wafer flatness, should be further improved. Elimination of sub-boundaries in the films also remains an important task, even though these defects do not seriously degrade MOSFET performance. We anticipate that the zone-melting recrystallization process can be developed sufficiently to yield SOI materials for VLSI applications.

Acknowledgements

This work was performed in collaboration with J. C. C. Fan, M. W. Geis, C. K. Chen, D. J. Silversmith and R. W. Mountain. The work was sponsored by the Department of the Air Force and the Defense Advanced Research Projects Agency.

REFERENCES

1) *Laser and Electron Beam Processing of Materials*, ed. C. W. White and P. S. Peercy (Academic Press, New York, 1980).
2) *Laser and Electron Beam-Solid Interactions and Material Processing*, ed. J. F. Gibbons, L. D. Hess, and T. W. Sigmon (North Holland, New York, 1981).
3) *Laser and Electron Beam Interactions with Solids*, ed. B. R. Appleton and G. K. Celler (North Holland, New York, 1982).
4) *Laser-Solid Interactions and Transient Thermal Processing of Materials*, ed. J. Narayan, W. L. Brown, and R. A. Lemons (North Holland, New York, 1983).
5) J. C. C. Fan, M. W. Geis, and B-Y. Tsaur: Appl. Phys. Lett. **38** (1981) 365.
6) E. W. Maby, M. W. Geis, Y. L. LeCoz, D. J. Silversmith, R. W. Mountain, and D. A. Antoniadis: IEEE Electron Dev. Lett. **EDL-2** (1981) 241.
7) M. W. Geis, H. I. Smith, B-Y. Tsaur, J. C. C. Fan, E. W. Maby, and D. A. Antoniadis: Appl. Phys. Lett. **40** (1982) 158.
8) A. Kamgar, G. A. Rozgonyi, and R. Knoell: Ref. 4, p. 569.
9) T. Stultz, J. Sturm, and J. F. Gibbons: Ref. 4, p. 463.
10) J. C. C. Fan, B-Y. Tsaur, R. L. Chapman, and M. W. Geis: Appl. Phys. Lett. **41** (1982) 186.
11) H. J. Leamy: Ref. 3, p. 459.
12) T. I. Kamins: J. Electrochem. Soc. **128** (1981) 1824.
13) J. C. C. Fan, B-Y. Tsaur, and M. W. Geis: J. Cryst. Growth **63** (1983) 453.
14) M. W. Geis, H. I. Smith, B-Y. Tsaur, J. C. C. Fan, D. J. Silversmith, and R. W. Mountain: J. Electrochem. Soc. **129** (1982) 2812.
15) K. A. Bezjian, H. I. Smith, J. M. Carter, and M. W. Geis: J. Electrochem. Soc. **129** (1982) 1848.
16) R. A. Lemons and M. A. Bosch: Appl. Phys. Lett. **40** (1982) 166.
17) W. G. Hawkins and D. K. Beigelsen: Appl. Phys. Lett. **42** (1983) 358.
18) B-Y. Tsaur, J. C. C. Fan, and M. W. Geis: Appl. Phys. Lett. **39** (1982) 561.
19) H. A. Atwater, H. I. Smith, and M. W. Geis: Appl. Phys. Lett. **41** (1982) 747.
20) R. F. Pinizzotto, H. W. Lam, and B. L. Vaandrager: Appl. Phys. Lett. **40** (1982) 388.
21) K. A. Jackson and C. E. Miller: J. Cryst. Growth **42** (1977) 364.
22) M. W. Geis, H. I. Smith, D. J. Silversmith, and R. W. Mountain: J. Electrochem. Soc. **130** (1983) 1178.
23) J. P. Colinge, E. Demoulin, D. Bensahel, and G. Auvert: Appl. Phys. Lett. **41** (1982) 346.
24) C. K. Chen, M. W. Geis, B-Y. Tsaur, and J. C. C. Fan: paper presented at the Electronic Materials Conference, Burlington, Vermont (June 22–24, 1983).
25) B-Y. Tsaur, M. W. Geis, J. C. C. Fan, D. J. Silversmith, and R. W. Mountain: Appl. Phys. Lett. **39** (1981) 909.
26) B-Y. Tsaur, J. C. C. Fan, M. W. Geis, D. J. Silversmith, and R. W. Mountain: IEDM Tech. Digest (1981) 232.
27) B-Y. Tsaur, J. C. C. Fan, M. W. Geis, D. J. Silversmith, and R. W. Mountain: IEEE Electron Dev. Lett. **EDL-3** (1982) 79.
28) B-Y. Tsaur, M. W. Geis, J. C. C. Fan, D. J. Silversmith, and R. W. Mountain: Ref. 3, p. 585.
29) B-Y. Tsaur, J. C. C. Fan, and M. W. Geis: Appl. Phys. Lett. **41** (1982) 83.
30) B-Y. Tsaur, J. C. C. Fan, R. L. Chapman, M. W. Geis, D. J. Silversmith, and R. W. Mountain: IEEE Electron Dev. Lett. **EDL-3** (1982) 398.
31) E. W. Maby and D. A. Antoniadis: Appl. Phys. Lett. **40** (1982) 691.
32) K. K. Ng, G. K. Celler, E. I. Povilonis, R. C. Frye, H. J. Leamy, and S. M. Sze: IEEE Electron Dev. Lett. **EDL-2** (1981) 316.

33) B-Y. Tsaur, J. C. C. Fan, G. W. Turner, and D. J. Silversmith: IEEE Electron Dev.
     Lett. **EDL-3** (1982) 195.
34) G. E. Davis, H. L. Hughes, and T. I. Kamins: IEEE Trans. Nucl. Sci. **NS-29** (1982) 1685.
35) See, for example, F. B. McLean: IEEE Trans. Nucl. Sci. **NS-27** (1980) 1651.
36) B-Y. Tsaur, R. W. Mountain, C. K. Cheu, G. W. Turner, and J. C. C. Fan: IEEE
     Electron Dev. Lett., July (1984)
37) G. Zimmer, B. Hoefflinger, and J. Schneider: IEEE Trans. Electron Dev. **ED-26** (1979)
     390.
38) S. Ogura, P. J. Tsang, W. W. Walker, D. L. Critchlow, and J. F. Shepard: IEDM
     Tech. Digest (1981) 651.
39) H. C. Lin, T. B. Tan, G. Y. Chang, B. Van der Leest, and N. Formigoni: *Proc.*
     IEEE **52** (1964) 1491.
40) A. S. Grove, *Physics and Technology of Semiconductor Devices* (John Wiley, New
     York, 1967) Chap. 7.

Silicon-on-Insulator: Its Technology and Applications, edited by S. Furukawa, pp. 129–136.
© KTK Scientific Publishers, Tokyo, 1985.

OPTICALLY-HEATED ZONE CRYSTAL GROWTH OF SILICON THIN FILMS ON AMORPHOUS SUBSTRATES

D. K. BIEGELSEN, W. G. HAWKINS,* L. E. FENNELL, N. M. JOHNSON, and M. D. MOYER

Xerox Palo Alto Research Center, Palo Alto, California 94304, U.S.A.

Abstract   In this paper we review the current understanding of issues relevant to the crystallization of silicon thin films on amorphous substrates. We treat in particular the case of radiant heating (*e.g.* lasers, lamps, strip-heaters, etc.). Semiconducting silicon becomes metallic and more highly reflecting on melting. Therefore, there is a natural negative feedback mechanism associated with optical coupling to the film. This results in inhomogeneous melting, which in turn leads to several very beneficial, as well as mildly deleterious consequences for crystal growth. On the positive side, the melting process is stabilized (robust) and texturing (specific crystal planes lying parallel to the substrate) occurs readily. On the other hand, undercooling is prevalent and can lead to growth instabilities. We describe a model for the phenomena of texturing and in-plane axial orientation. We then discuss mechanisms of defect formation (*e.g.* low-angle grain boundaries, twinning—and in the extreme—amorphous solidification). Finally, we discuss methods for controlling the lateral epitaxial growth from single oriented seeds.

In the past few years much effort has focused on zone melting recrystallization of silicon thin films on amorphous insulators.[1] The most common energy sources have been radiant—*e.g.*, lasers, strip heaters, lamps—as opposed to nonradiant—*e.g.*, electron beams or ohmic electron currents. Swept zone lateral epitaxy has resulted in material of surprisingly good crystalline and electronic quality. Our understanding of the basic mechanisms and our ability to control the process to achieve reliable results are both improving rapidly. In this paper we review the current state of lateral epi in radiantly-heated silicon thin films with particular emphasis placed on bulk amorphous substrates.

Both conceptually and operationally the thin film crystal growth pro-

*Xerox Webster Research Center, Webster, N.Y. 14580, U.S.A. .

cess can be divided into two stages: autonucleation and later epitaxial growth. We refer to autonucleation here because it has been found that one can achieve almost equivalent results (with greater flexibility) if no pre-existing seed is used, such as a via to a crystalline substrate. We therefore discuss first the physics of the initiation of crystal growth.

Consider the case of uniform, strongly-absorbed illumination of a silicon film on a heat-sunk substrate. Figure 1 shows schematically the inrease in temperature of the solid film as the incident flux is increased. Liquid silicon is metallic and therefore both more strongly absorbing and reflecting than the hot solid. For the low translational velocities normally used in zone melting, the sample temperature is near the melting temperature under most of the irradiated region. We find then that the absorption length is short compared to the sample thickness both in the melt and hot solid; thus, all light entering the film is totally absorbed. It is therefore the change in reflectivity that controls the power absorption. As indicated in Fig. 1, it takes approximately twice the power to maintain a molten region at T_m than for a solid region; $P_2 \sim 2 \cdot P_1$. This is a natural negative feedback situation;[2] for a flux P between P_1 and P_2 the sample responds by separating into coexisting liquid and solid regions[3] (lamellae, Fig. 2) so as to maintain a constant absorbed power and, hence, a constant temperature, T_m. The characteristic size of the lamellae is determined by simultaneously satisfying two countervailing tendencies in the film. In brief, these are (1), the increased undercooling of the liquid regions as their size grows and (2), the tendency to nucleate a solid phase in the coolest

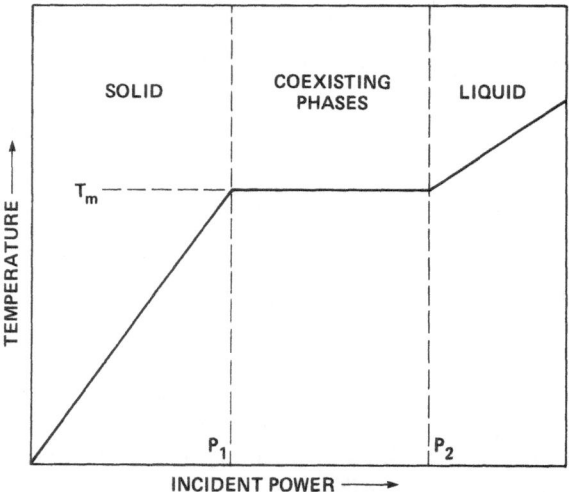

Fig. 1.   Film temperature versus incident power. P_1 and P_2 denote the onset of inhomogeneous and total melting, respectively.

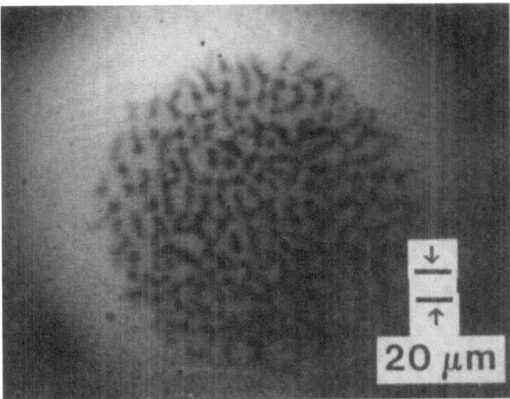

Fig. 2. Video image of blackbody radiation from inhomogeneously melted silicon thin film on silica substrate.

regions of the undercooled liquid. (The argument is totally invertable for superheated solid regions.)

It has been shown[4] that for wavelengths less than about five times the lamellar widths, it is correct to use a short wavelength (geometric) model for the optical coupling (as opposed to an effective medium treatment.) That is, the light is absorbed in accordance with the local optical constants of the material. For a model based on alternating stripes of liquid and solid it has been shown[2] that the undercooling at the center of the liquid region, ΔT, increases monotonically as the liquid width increases. This is plotted in Fig. 3 and labeled "heat flow" because the degree of undercooling depends on the thermal properties of the film and substrate. For the second part of the problem we must write down an expression for the change in free energy, ΔE, if the center of the undercooled liquid is replaced by solid. There are three dominant contributions—a bulk energy difference favoring creation of the low temperature phase, and two interfacial energy differences both of opposite sign. If we assume lamellar stripes of width w, height h and length l, then

$$\Delta E = -(whl)L\Delta T + (2wl)\Delta\gamma_{sub} + (2hl)\gamma_{SL}$$

where L is the latent heat of melting, γ_{sub} is the specific interfacial free energy between the silicon lamellae and the SiO_2 layers below and above (although encapsulation is not necessary to this argument), and γ_{SL} is the specific solid-liquid interfacial energy. Both interfacial energies are expected to be large and axially anisotropic for silicon.[5] If ΔE is greater than zero, the fluctuation into the solid phase is unstable and the solid remelts. The stability condition, $\Delta E = 0$, is plotted in Fig. 3 and labeled "nucleation."

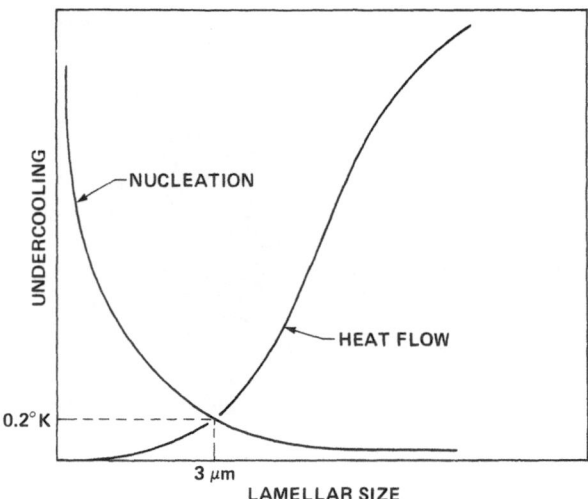

Fig. 3. Mechanisms determining the size of lamellae. Liquid undercooling increases with lamellar width; minimum stable size of solid inclusion decreases with liquid undercooling.

The intersection of the two curves represents the physically allowed state— i.e. the characteristic size and maximum undercooling of the liquid lamellae. For 0.5 μm silicon films on bulk SiO$_2$, typical resulting values are ~3 μm and 0.2°K respectively. The steady state parameters clearly depend on the thermal properties of the substrate and are quite different, for example, for a crystal silicon substrate held at room temperature.

The steady state inhomogeneous melting phenomenon leads to a particularly beneficial effect. Because undercooling is small, so that only large solid regions are stable, and because the domains exist for times long compared with crystal regrowth times, the interfacial terms are relatively strong and the crystalline lamellae are able to minimize their energy by growing with an optimum axial orientation. Experiments are in progress to determine if it is γ_{sub} or γ_{SL} which dominates the anisotropic nucleation. Experiments performed at room temperature, i.e. after cool down,[6,7] have demonstrated that there is a strong preferential orientation with {100} planes parallel to the SiO$_2$ interfaces. Also studies of graphoepitaxy have shown that the ordering occurs within the coexistence phase and depend on the substrate material. Thus, γ_{sub} is strongly indicated. Recently Raman scattering has been used to probe *in situ* the structure of the solid lamellae.[8] It was shown that the lamellae indeed are crystalline. (This, by the way is the first direct proof that the amorphous structure has higher energy than the crystal at T_m.) Because of the difficulty of the experiments, there is not direct proof yet of a common crystal axis normal to the substrate. The {100} lamellae have random in-plane orientation. However, when the zone is moved, kinetic effects lead to a selection of a small distribution of directions relative to the thermal gradients which grow most rapidly and dominate the lateral epitaxial growth.[7] In the laser-induced zone melting, crystallites with the $\langle 220 \rangle$ axis (trace of the $\{111\}$ plane) lying parallel to the maximum thermal gradient predominate. In situations having smaller thermal gradients, due for example to substrate bias heating (e.g. strip heater ovens), the $\langle 100 \rangle$ directions align with the thermal gradient and a faceted growth front is observed. We find then that in sweeping a radiantly-heated molten zone through a thin silicon film, the epitaxial wake tends to consist of parallel grains of $\{100\}$ textured material separated by low angle grain boundaries.

Before proceeding into a discussion of continued growth from these autonucleated seeds, we wish to point out how some self-limiting behaviors arise from fundamental properties and how they lead to both positive and negative ramifications for crystal growth. It is already obvious that the inhomogeneous melting leads to an intensity-independent film temperature for power variations of a factor of two. But even in situations where no lamellae form, the high thermal diffusivity of molten silicon gives rise to a temperature flattening mechanism for silicon films on insulating substrates. Consider, for example, the situation with a thick optically absorbing encapsulation layer. Then the light is absorbed in the encapsulant and the silicon is heated indirectly by conduction so that no optical feedback results. However, the high thermal diffusivity (relative to the solid silicon and substrate) within the molten zone means that the melt tends to be isothermal. The temperature is pinned to T_m at the liquid-solid interface. Fluctuations in the incident flux result primarily in modulation of the zone width, but only secondarily affect the temperature.

Both these mechanisms greatly reduce the sensitivity of the growth process to source instabilities. On the other hand, both mechanisms also result in shallow positive to strongly negative thermal gradients into the melt. These conditions are conducive to growth interface instability. Indeed, the defects encountered in these systems (e.g. twinning, filamentary growth, etc.) are characteristic of a strongly undercooled melt.[9] We can roughly categorize three regimes of defect formation corresponding to the degree of undercooling. For fairly flat (positive or negative) thermal gradients, faceting tends to develop in the growth interface and low angle grain boundaries are nucleated and propagated. For high degrees of supercooling, occurring for example when a downstream region is penninsular[10] or for large fluctuations in growth velocity, random nucleation occurs from very small clusters and growth is rapid and dendritic. Finally, at exceedingly large undercooling (several hundred degrees) typical of sub-nanosecond pulsed melting and cooling times, amorphous material can be quenched in.[11]

The role played by the encapsulating layer is not yet fully understood. Intentional encapsulation was used initially[12,13] to reduce dewetting and minimize mass transport effects on topography. Subsequently it was shown that SiO_2 encapsulants also enhanced texturing.[14] It has been suggested that the encapsulant acts as a stretched membrane over the lamellae and could lead to an orientation-dependent stress-induced elastic energy contribution to the solid free energy. However, {100} texturing occurs even without deposited encapsulants and often does not occur for Si_3N_4 encapsulation. We therefore believe that control of the interfacial energy is the dominant mechanism in crystal orientation.

An intrinsic mechanism for controllable oriented nucleation thus exists. We now turn to the issues of stable lateral epitaxial growth. The conditions which optimize the oriented nucleation process are not necessarily the same as those which lead to best continued crystal propagation. For example, the nucleation mechanism requires optical absorption in the silicon as a primary heat source. This implies optically thin encapsulating layers and seeding regions having lateral dimensions greater than a thermal diffusion length, d_T. That is, increased absorption in the substrate or overlayers can conduct into the film within d_T and negate the possibility of undercooling in liquid regions. On the other hand, for lateral crystal growth, continuing nucleation competing with the epitaxy is undesirable. Furthermore, achieving flat interface contours and positive thermal gradients is important. Using optically absorbing layers or silicon pattern formation one can control the lateral thermal profile of the zone.[13] Using patterned regions less than d_T or optically thick encapsulants leads to removal of lamellar formation and liquid supercooling. Figure 4 indicates schematically a few of the techniques which have been used to create desirable profiles. Figure 4A indicates a self-aligning method to convert a convex irradiation profile into a concave thermal profile.[15] It utilizes a moat pattern in the silicon film. Infrared radiation ($\lambda > 5$ μm) absorbed in the substrate causes a temperature rise which is not limited at T_m. Thus edge heating ensues. Figure 4B shows the result of patterning when visible radiation is used and the substrate is an oxidized silicon wafer. Here the oxide increases the optical coupling into the substrate and again leads to edge heating. In Fig. 4C we show a variation pursued by Colinge *et al.*[16] A continuous silicon film has overlayer rails which enhance the optical coupling at the sides. In Figs. 4A–C polycrystalline growth occurs outside the thermal peaks. Geis *et al.*[17] have used a similar technique with a line source to create lateral thermal gradients for defect sinking. Figure 4D represents spot shaping to control the optical absorption.[18, 19] Figure 4E then shows the type of thermal profile one can achieve by these methods. The process parameters can easily be modified to optimize the growth interface profile. The variety of alternatives is indicative of the flexibility afforded by

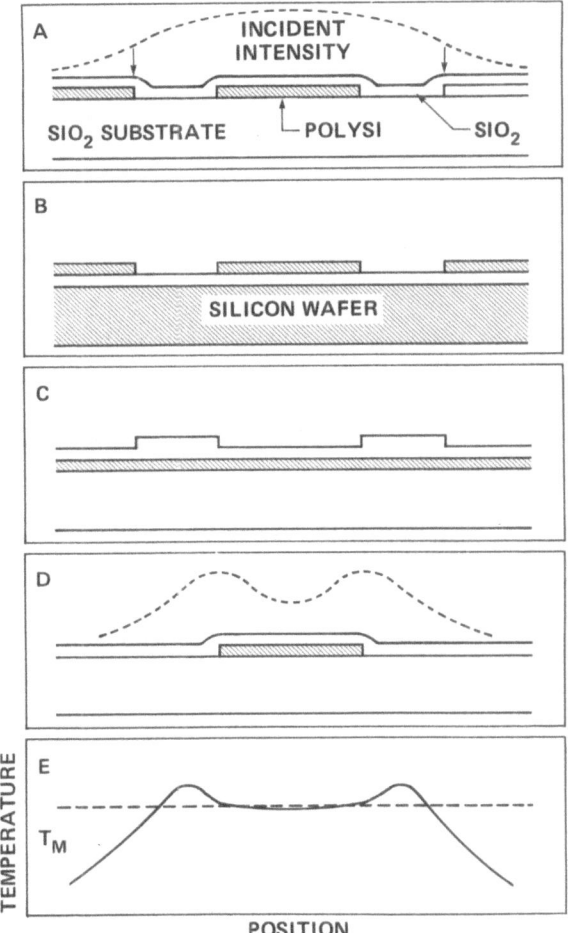

Fig. 4. (A) Patterned polysilicon film on silica substrate. Illumination spot is larger than moat separation. (B) Patterned poly on oxidized silicon wafer. (C) Encapsulation layer thickness modulatd to affect optical absorption. (D) Spot shaping to compensate for edge cooling with visible radiation. (E) Typical thermal profile expected for each of the above configurations.

surface access.

Patterning provides further benefits. Atwater et. al.[20] have used an "hourglass" delineation to pinch off sub-boundary propagation. In a similar way we transition from a continuous polynucleation region into stripes for continued growth. Only one of a multiplicity of low angle related seeds is allowed into each stripe.

The field of zone growth in thin films on amorphous substrates is rapidly coming to fruition. Control of oriented crystal growth has been demonstrated on a laboratory scale and high yields of excellent thin film transistors have been achieved. Nevertheless there exists a great deal of fundamental research to be done to fill out our understanding of crystal growth in regimes where interfaces are so important. To be able to achieve a viable technology, issues such as high growth rates and steeper vertical thermal gradients must be addressed. There are undoubtedly many surprises still ahead.

REFERENCES

1) See Proceedings of MRS Symposia (Elsevier, New York).
2) W. G. Hawkins and D. K. Biegelsen: Appl. Phys. Lett. **42** (1983) 358.
3) M. A. Bosch and R. A. Lemons: Phys. Rev. Lett. **47** (1981) 1151.
4) D. Aspnes: private communication.
5) See, for example, Yu. V. Naidich, N. F. Grigorenko, and V. M. Perevertailo: J. Cryst. Growth **53** (1981) 261.
6) M. W. Geis, H. I. Smith, B.-Y. Tsaur, J. C. C. Fan, D. J. Silversmith, and R. W. Mountain: J. Electrochem. Soc. **129** (1982) 2812.
7) D. K. Biegelsen, N. M. Johnson, W. G. Hawkins, L. E. Fennell, and M. D. Moyer: *Laser Solid Interactions and Transient Thermal Processing of Materials*, ed. J. Narayan, W. L. Brown, and R. A. Lemons (Elsevier, New York, 1983).
8) D. K. Biegelsen, R. J. Nemanich, and R. A. Street (to be published).
9) W. A. Tiller: *Art and Science of Growing Crystals*, ed. J. J. Gilman (J. Wiley, New York, 1964) p. 310.
10) J. G. Black, W. G. Hawkins, and C. H. Griffiths: J. Appl. Phys. (to be published).
11) A. G. Cullis, H. C. Webber, and N. G. Chew: Appl. Phys. Lett. **42** (1983) 875.
12) A. R. Billings: J. Vac.Sci. & Technol. **6** (1969) 757.
13) D. K. Biegelsen, N. M. Johnson, D. J. Bartelink, and M. D. Moyer: Appl. Phys. Lett. **38** (1981) 150.
14) M. W. Geis, D. A. Antoniadis, D. J. Silversmith, R. W. Mountain, and H. I. Smith: Appl. Phys. Lett. **37** (1980) 454.
15) W. G. Hawkins, J. G. Black, and C. H. Griffiths: Appl. Phys. Lett. **40** (1982) 319.
16) J. P. Colinge, E. Demoulin, D. Bensahel, G. Auvert, and H. Morel: IEEE Electron Device Lett. **4** (1983) 75.
17) M. W. Geis, H. I. Smith, D. J. Silversmith, R. W. Mountain, and C. V. Thompson: J. Electrochem. Soc. (in press).
18) S. Kawamura, J. Sakurai, M. Nakano, and M. Takagi: Appl. Phys. Lett. **40** (1982) 394.
19) D. K. Biegelsen, N. M. Johnson, D. J. Bartelink, and M. D. Moyer: *Laser and Electron Beam Interactions*, ed. J. F. Gibbons, L. D. Hess, and T. W. Sigmon (Elsevier, New York, 1981) p. 487.
20) H. A. Atwater, H. I. Smith, and M. W. Geis: Appl. Phys. Lett. **41** (1982) 747.

Silicon-on-Insulator: Its Technology and Applications, edited by S. Furukawa, pp. 137–150.
© KTK Scientific Publishers, Tokyo, 1985.

RECRYSTALLIZATION OF POLYCRYSTALLINE SILICON ON FUSED SILICA USING AN RF-HEATED CARBON SUSCEPTOR

Y. KOBAYASHI, A. FUKAMI, and T. SUZUKI

Hitachi Research Laboratory, HITACHI, Ltd. Hitachi-shi, Ibaraki 319-12, Japan

Abstract A new zone melting recrystallization method that uses an RF-heated carbon susceptor in fabrication of SOI (silicon on insulator) structures has been developed. In this method, a 0.5 to 1.0 μm thick polycrystalline silicon film, encapsulated with a 1.2 μm thick CVD-SiO$_2$ layer, was deposited on a fused silica substrate. The substrate was moved across the carbon susceptor, which had a narrow, high temperature zone. The continuous silicon films produced had grains in the recrystallized silicon film of several tenths to a few mm wide and a few cm long, and a film orientation of (100). In the grains, there were many small angle grain boundaries which consisted of discrete dislocations. There were some cracks in the silicon film. In the silicon islands, recrystallized silicon was a single crystal with an orientation of (111) and there were no cracks. In order to obtain silicon without cracks and with an orientation of (100), a method that connects the polycrystalline silicon islands was proposed.

1. Introduction

Silicon on insulator(SOI) structures offers many advantages for display devices such as active liquid crystal displays, for high speed integrated circuits and for three dimensional circuits. The structures have been produced by melt regrowth methods using laser beams,[1,2] electron beams[3,4,16] and incoherent energy sources such as strip carbon heaters[5-7] and arc-lamps.[8,9] The laser and electron beam recrystallization methods have the potential to produce three dimensional circuits, but have a small throughput. On the other hand, zone melting recrystallization using incoherent energy sources is not suitable for three dimensional circuits because the entire wafer must be heated. However, a high quality silicon layer can be obtained, without seeding, as can high throughput.[5] Therefore, this method is suitable for manufacturing very high speed circuits and display devices.

In the zone melting recrystallization method, temperature control in the melted silicon region and its proper distribution is necessary to prevent agglomeration of the melted silicon and to obtain a good surface uniformity.

The authors have developed a novel zone melting regrowth method using an RF-heated carbon susceptor, in which temperature and shape of the melted zone can be easily observed and controlled.[10] This article describes the method and the results obtained while investigating the quality of the recrystallized films produced.

2. RF-Heated Zone Melting Regrowth Method (RF-ZMR Method)

Figure 1 shows the recrystallization method. The upper figure (a) is a cross sectional view of the apparatus, and the lower figure (b) is a schematic blow-up of the wafer and carbon susceptor area. A 0.5 to 1.0 μm thick polycrystalline silicon film was deposited on an optically polished, fused silica substrate. The silicon surface was covered with a 1.2 μm

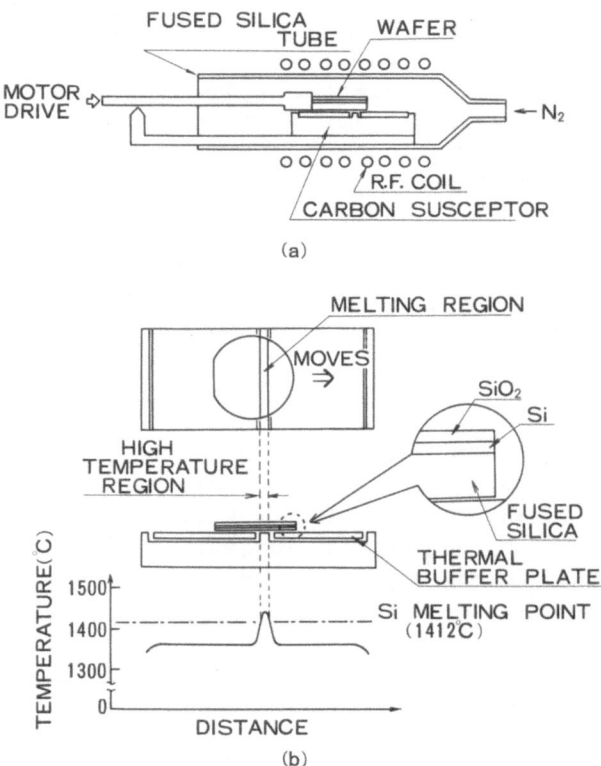

Fig. 1.   (a) Schematic diagram and (b) illustration of the zone melting recrystallization method.

thick layer of silicon dioxide by a vapor deposition process. The wafer
was then moved across the carbon susceptor surface (speed: 0.5 or 1 mm/s),
which was RF-heated and had the temperature distribution reproduced
in Fig. 1 (b). In the high temperature region (1 or 2 mm wide), where
the polycrystalline silicon was melted, a higher heat flow from the RF-
heated carbon susceptor came into contact with samples. In the low
temperature regions, the heat flow was partially shielded by fused silica
or carbon thermal buffer plates. The temperature of the high temperature
region, measured using a pyrometer, was about 1450°C ; 100°C higher
than that of other regions. Melting and regrowth took place in a nitrogen
atmosphere. The silicon layer was melted in the high temperature region
and then, after moving on, was solidified to form a regrown layer.

Figure 2 is a photograph, taken during recrystallization, of the melted
silicon. An infrared transmitting glass filter was used. Dark zones were
shadows produced by the work coil. The circular region was the wafer
(diameter:50 mm). The light colored strip was the high temperature region,
while the slightly darker, narrow ellipse was the melted silicon region.
The area of the melted silicon was probably darker than that of the surrounding unmelted silicon due to the lower emissivity of the melted silicon
as compared to unmelted silicon. The photograph demonstrated that the
melted silicon could be easily observed and that recrystallization conditions could be easily controlled.

3. Characteristics of the Recrystallized Silicon

Figure 3 shows (a) an optical micrograph of an etched silicon surface

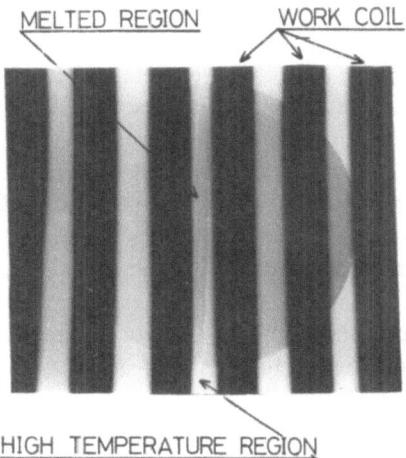

Fig. 2.   Photograph of the melted silicon region during recrystallization.

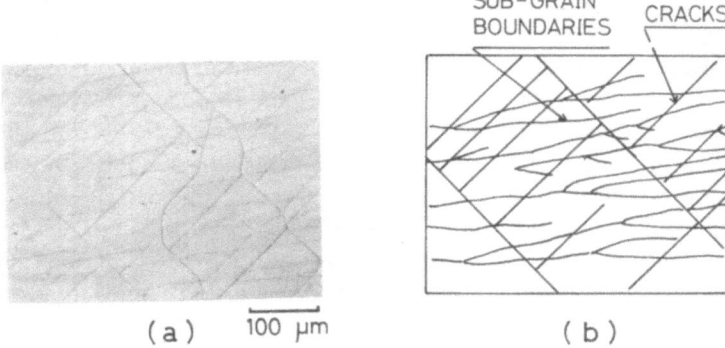

Fig. 3.   (a) Optical micrograph of an etched silicon surface after recrystallization. (b) Schematic of the optical microgaph.

after recrystallization and (b) its schematic. The etchant composition was $HF:HNO_3:CH_3COOH = 1:50:50$[5]. There were many elongated sub-grains and cracks. The sizes of the sub-grains were several tens of μm wide and several hundreds of μm long. All cracks were parallel or perpendicular. These crack patterns meant; that the orientation of the recrystallized silicon was (100), because the direction of the cracks were ⟨110⟩ for the single crystal silicon; and that the sub-grain boundaries were small angle grain boundaries, because the cracks were straight even when they crossed the sub-grain boundaries. These points were studied in detail with transmission electron microscopy.

Figure 4 is a transmission electron micrograph (TEM) of the recrystallized silicon. Three diffraction patterns which correspond to the indicated region are also shown. Dislocations and sub-grain boundaries could be observed. The diffraction patterns showed that orientations of the three regions were (100). All the patterns had the same rotation angle within one degree or less. So, it was concluded that the sub-grain boundaries were small angle grain boundaries.

Figure 5 is a magnified photograph of one diffraction pattern. The Kikuchi lines and bands could be clearly observed, indicating that the crystalline quality of the recrystallized silicon was high. Results of the TEM observations corresponded to those observed for the etched silicon surface.

Next, the sub-grain boundary was studied using TEM. A sub-grain boundary has been assumed to consist of an array of dislocation.[5,7] Figure 6 is a 1000 keV TEM photograph of a sub-grain boundary. Individual dislocations could be seen on the sub-grain boundary ; that is, the boundaries consisted of discrete dislocations. When the deviation of the orientation was estimated from the distance between these dislocations, a value

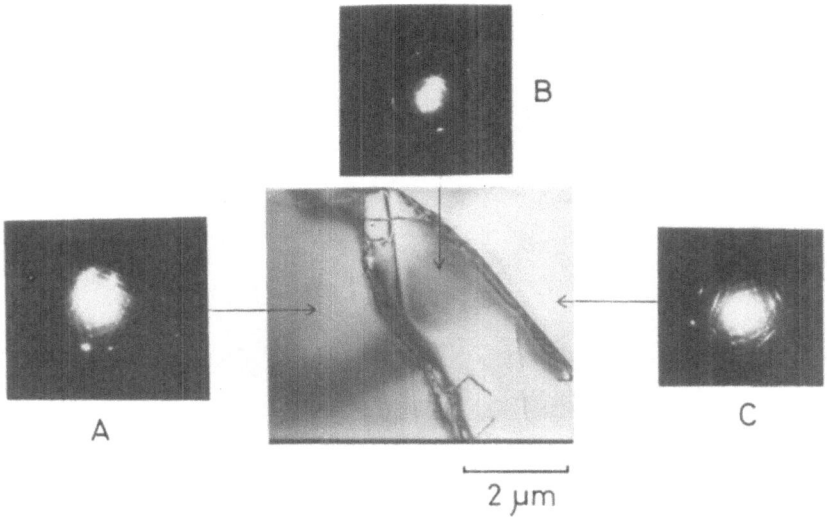

Fig. 4. Transmission electron micrograph of a recrystallized silicon layer, and diffraction patterns from each sub-grains area.

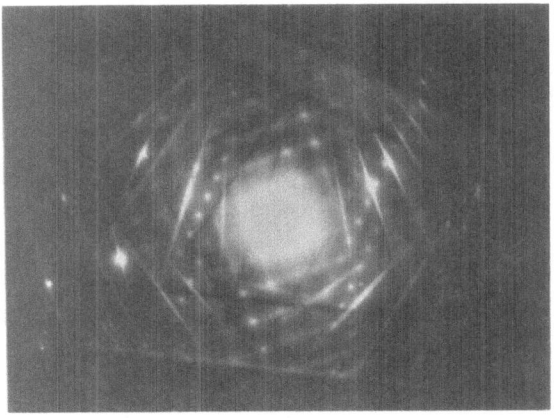

Fig. 5. Magnified photograph of a diffraction pattern.

of about 0.6 deg. was obtained.

As shown in Fig. 7, an Al-gate, n-channel MOSFET was fabricated on the recrystallized silicon, in order to investigate its electrical properties. The gate length and gate width were 20 and 723 μm, respectively. The thickness of the gate oxide film was 1000 Å.

Figure 8 shows electrical characteristics of (a) a MOSFET fabricated

Y. Kobayashi *et al.*

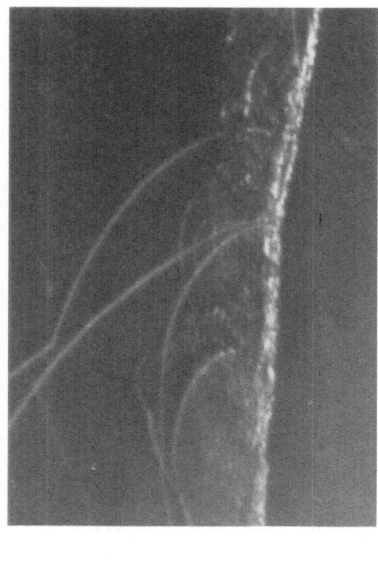

0.5 μm

Fig. 6. 1000 kV transmission electron micrograph of a sub-grain boundary.

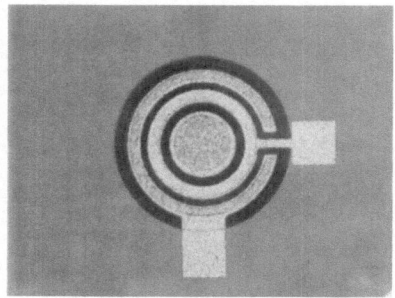

DIMENSIONS

GATE LENGTH	:	20 μm
GATE WIDTH	:	723 μm
GATE FILM THICKNESS	:	1000 Å

Fig. 7. Photograph of a MOSFET fablicated from a recrystallized siliocn.

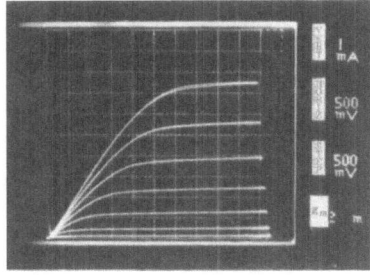   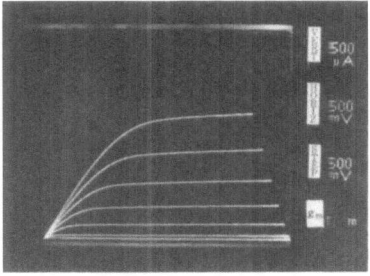

SOI/MOSFET   BULK Si/MOSFET
(μ = 1023 cm^2/ V · s)   (μ = 569 cm^2/ V · s)

Fig. 8.   Electrical characteristics of MOSFET faricated from (a) recrystallized silicon and (b) single crystal bulk silicon.

on recrystallized silicon and (b) a MOSFET fabricated on bulk, single crystal silicon using the same process as the SOI/MOSFET fabrication. Electron mobility could be obtained from these characteristics. The maximum electron mobility in the saturated region of the recrystallized silicon was about 1020 cm^2/Vs, while the average value was about 860 cm^2/Vs. On the other hand, the electron mobility of the recrystallized silicon was about 600 cm^2/Vs. That is, the mobility of the recrystallized silicon was about 1.4 times that of the bulk silicon. One of the reasons for the higher electron mobility observed in the recrystallized silicon was tensile stress in the silicon due to a difference in the thermal expansion coefficients of fused silica substrate and silicon. The tensile stress in a (100) silicon plane reduced the effective mass of an electron, which increased electron mobility.[6,11]

These experimental results described above verified that this RF-ZMR method would produce a high quality recrystallized silicon on a fused silica substrate.

4.   Recrystallization of Polycrystalline Silicon Islands

In structures such as silicon on fused silica, the serious problem is cracking in the recrystallized silicon caused by stress,[12,13] as previously shown in Fig. 3. Figure 9 (a) is a photograph of the recrystallized silicon surface having some cracks. And Figure 9 (b) shows the electrical characteristics of a MOSFET fabricated from the cracked silicon. The characteristics of the devices were abnormal, indicating the necessity of eliminating cracks in the recrystallized silion. Therefore, reduction of stress

CRACK

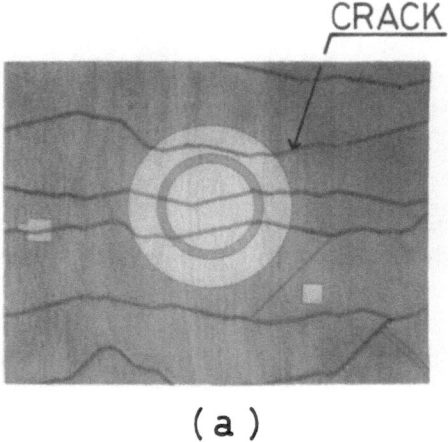

(a)

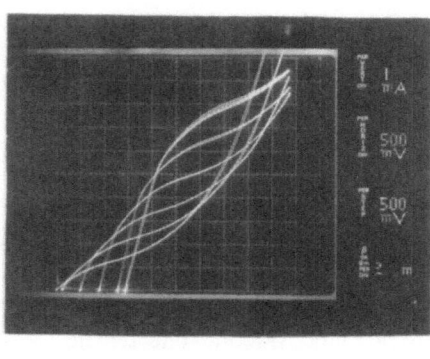

(b)

Fig. 9. (a) Photograph of a recrystallized silicon surface. (b) Electrical characteristics of a MOSFET having cracks.

by forming silicon islands was tried.

The polycrystalline silicon film was then etched with a dry etching technique to form islands with areas of from $5 \times 5 \ \mu m^2$ to $4 \times 4 \ mm^2$. Subsequent processes were the same as for the continuous silicon described in Section 3. The crack density of the recrystallized silicon was measured after the encapsulating SiO_2 layer was etched off.

As shown in Fig. 10, cracks disappeared when recrystallized islands of polycrystalline silicon were less than $250 \times 500 \ \mu m^2$ in size when silicon thickness was 0.5 μm, and when islands were less than $50 \times 100 \ \mu m^2$ in size when the silicon thickness was 1 μm.

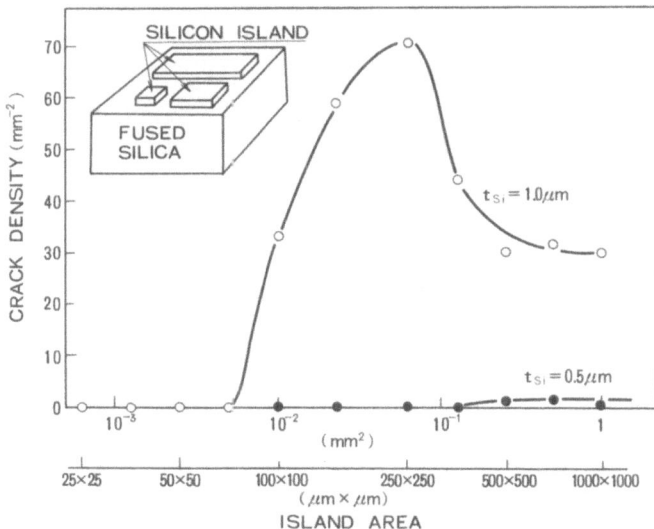

Fig. 10.   Relationship between crack density and island area.

Figure 11 shows a SEM photograph of the Sirtl[14] etched silicon surface. The lines of dark dots were etch pits, corresponding to dislocations. These indicated that recrystallized silicon islands had some subgrain boundaries consisting of arrays of discrete dislocations, but no grain boundaries. This absence indicated that the recrystallized silicon in an island was a single crystal. The orientation of the recrystallized silicon islands was observed with a TEM technique. Figure 12 shows a typical diffraction pattern observed for one island. Orientation of the silicon island was (111). Figure 13 shows an optical micrograph of the silicon island surface which was wider than that for which no cracks occurred. As shown in this figure, there were many cracks and their arrangements were equilateral triangles.

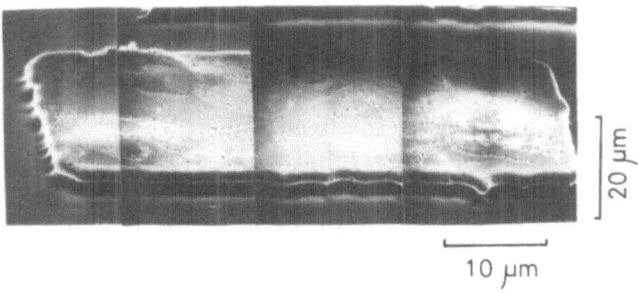

Fig. 11.   SEM photograph of Sirtl etched surface of recrystallized silicon island.

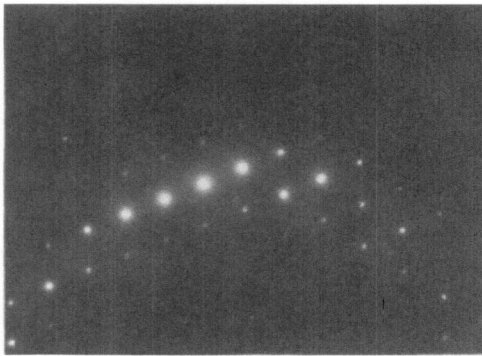

Fig. 12.   A typical diffraction pattern of the recrystallized silicon.

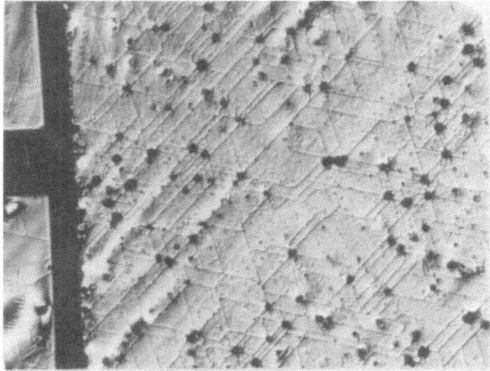

Fig. 13.   Crack shapes in a silicon island which is wider than that for which no cracks occurred.

This meant that the orientation of the island was (111), because a crack line was ⟨110⟩. Orientation of all another islands were justified by the arrangements of the cracks, and were (111).

   In order to study the electrical properties of the recrystallized silicon islands, ring type, Al-gate, and n-channel MOSFETs were fabricated on them. The thickness of the gate oxide film was 1200 Å. Gate length and width were 20 and 680 μm, respectively. Figure 14 shows the field effect mobilities of devices with a drain voltage of 0.1 V. These mobilities depended on the the threshold voltages of the devices. The reference MOSFETs were also fabricated in bulk silicon with an orientation of (111) using the same fabrication process. The field effect mobility of a device on the recrystallized silicon island was the same as that of a device on bulk silicon with an orientation of (111) for the same device threshould voltages.

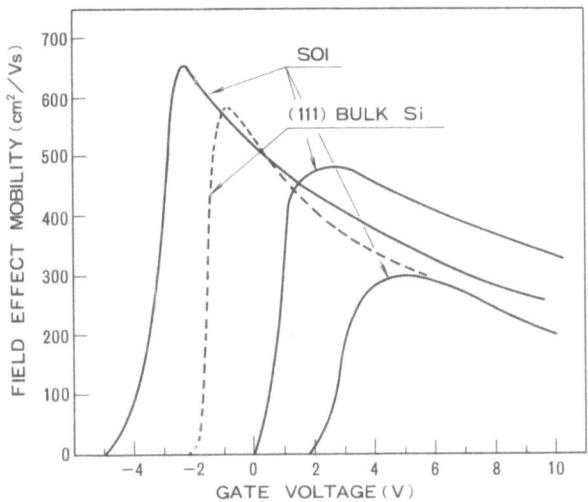

Fig. 14. Field effect mobility of MOSFETs fabricated on recrystallized silicon.

Leakage current was also measured at a drain voltage of 5 V, and found to be 1.5×10^{-11} A/μm.

As described above, cracks were eliminated by recrystallization of the island polycrystalline silicon on fused silica substrate. The recrystallized silicon island was a single crystal having an orientation of (111), ensuring good electrical properties. However, tensile stress could not increase the electron mobility, such as in recrystallization of a continuous polycrystalline silicon film, because of the island orientation. Consequently, it was necessary to obtain (100) oriented silicon by recrystallizing in order to produce a high speed switching a MOSFET. The difference in orientation of continuous and island silicon is discussed next.

Figure 15 illustrates the orientation difference of continuous and island silicon. In the continuous silicon, heat flowed smoothly in a direction parallel to the surface of the silicon because the silicon was sandwiched between the SiO$_2$ film and the fused silica substrate, which both had thermal conductivity coefficients two orders lower than that of silicon. On the other hand, in the island silicon, heat was held in the melted silicon region because the silicon was completely enclosed by the SiO$_2$ layer and the fused silica. So, the difference in orientation may be caused by a difference in heat flow during silicon recrystallization. Supercooling of the melted silicon was observed to a temperature more than 200°C less than the melting point of silicon, supporting the above explanation.

Based on this explanation, a method for obtaining crackless and (100) oriented silicon on fused silica was proposed. As shown in Fig. 16, silicon

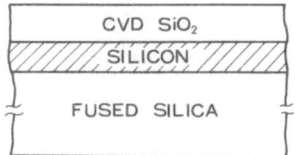

A) IN CASE OF CONTINUOUS SILICON

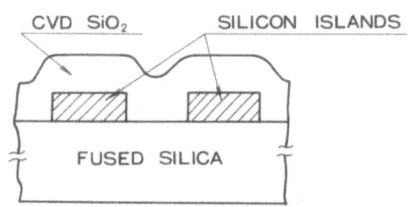

B) IN CASE OE SILICON ISLANDS

Fig. 15.   Illustrations of the orientation difference between recrystallization (a) of a continuous silicon layer and (b) of silicon islnds.

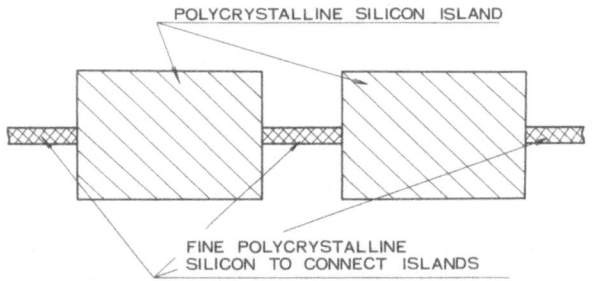

Fig. 16.   Illustration of a connected silicon island pattern.

was formed as an island in order to eliminate cracking and then islands were connected to each other with fine polycrystalline silicon in order to achieve a smooth flow heat. Then the polycrystlline silicon could be recrystallized.

Figure 17 shows the orientation of silicon recrystallized by this method. The orientation was studied with the etch pit method (figure insert)[15]. The shape of the etch pits was square, so orientation of the recrystallized silicon was (100) and rotation angles of the squares were all the same in an island, hence the island silicon was a single crystal.

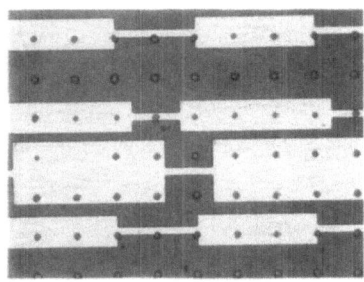 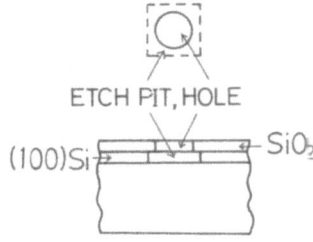

Fig. 17.   Orientation of the recrystallized silicon islands which are connected with fine silicon regions.

5. Conclusion

A new zone melting recrystallization method using an RF-heated carbon susceptor has been developed. This method allows easy control of the temperature and distribution of the melted silicon region, which is an important factor in zone melting recrystallization. In this method, it is necessary to recrystallize the connected polycrystalline silicon islands in order to obtain high quality crystalline silicon without cracks and with an orientation of (100).

Acknowledgements
The authors would like to express their sincere thanks to Prof. T. Imura and Dr. H. Saka for taking the 1000 kV transmission electron micrographs and their useful discussions regarding crystalline quality. They would also like to thank to Drs. T. Takasuna and M. Okamura for their guidance and encouragement.

REFERENCES

1) M. Tamura, M. Ohkura, and T. Tokuyama: Jpn. J. Appl. Phys. **21**, (1981). Suppl. 21-1.
2) H. W. Lam, Z. P. Sobczak, R. F. Pinizzotto, and A. F. Tasch Jr.: IEDM Tech. Dig. (Washington, DC) (1980) 559.
3) K. Shibata, T. Inoue, and T. Takigawa: Appl. Phys. Lett. **39** (1981) 645.
4) T. I. Kamins and B. P. Von Herzen: Electron Dev., Lett. **EDL-2** (1981) 313.
5) E. W. Maby, M. W. Geis, Y. L. LeCoz, D. J. Silversmith, R. W. Mountain, and D. A. Antoniadis: Electron Dev. Lett. **EDL-2** (1981) 241.
6) B-Y. Tsaur, John C. C. Fan, and M. W. Geis: Appl. Phys. Lett. **40** (1982) 322
7) R. F. Pinizzotto, H. W. Lam, and B. L. Vaandrager: Appl. Phys. Lett. **40** (1982) 388.
8) T. J. Stultz and J. F. Gibbons: Appl. Phys. Lett. **41** (1982) 824.
9) A. Kamgar and E. Labate: Mat. Lett. **1** (1982) 91; E. Siltl and A. Adler: Z. Metallkde. **52** (1961) 529.
10) Y. Kobayashi, A. Fukami, and T. Suzuki: Electron Dev. Lett. **EDL-4** (1983) 132.
11) Y. Kobayashi, M. Nakamura, and T. Suzuki: Appl. Phys. Lett. **40** (1982) 1040.

12) T. I. Kamins and P. A. Pianetta: Electron Dev. Lett. **EDL-1** (1980) 214.
13) W. G. Hawkins, J. G. Black, and C. H. Griffith: Appl. Phys. Lett. **40** (1982) 319.
14) E. Siltl and A. Adler: Z. Metallkde. **52** (1961) 529.
15) M. W. Geis, H. I. Smith, B-Y. Tsaur, J. C. C. Fan, E. E. May, and D. A. Antoniadis: Appl. Phys. Lett. **40** (1982) 158.
16) H. Ishiwara, M. Nakano, H. Yamamoto, and S. Furukawa: Jpn. J. Appl. Phys., **22**, (1983) Suppl 22-1, 607.

Silicon-on-Insulator: Its Technology and Applications, edited by S. Furukawa, pp. 151–157.
© KTK Scientific Publishers, Tokyo, 1985.

STRIP HEATER RECRYSTALLIZED SOI STRUCTURES

K. HIGUCHI, S. SAITOH, and H. OKABAYASHI

*Microelectronics Research Labs., NEC Corporation,
Miyazaki, Miyamae-ku, Kawasaki 213, Japan*

Abstract   Si films recrystallization on SiO_2 was performed using the strip heater method. The recrystallized Si films crystallinity was found to be dependent on the width of SOI islands surrounded by a seeding area, based on electron channeling pattern observations. SOI islands with less than 200 μm width were recrystallized into single crystals. In a few mm wide SOI islands, subgrain growth was observed, although each subgrain oriented in nearly seed crystal orientation. Large angle grains, which contained many subgrains, were grown for the sample with a seeding area at one side.

1. Introduction

Si films recrystallization on SiO_2, using the zone melting recrystallization by strip heater method, is characterized in preparation of large area SOI. However, such SOI films typically contain about 50 μm wide and several hundred μm long grains or subgrains.[1-3]

Some authors reported that subboundaries did not make fatal defects in SOI devices by measuring lifetimes and electrical MOSFETs characteristics in zone melting recrystallized Si films on SiO_2.[4-6] On the other hand, it was also reported that dopants in source and drain regions in MOSFETs extraordinarily diffused through grain boundaries and that channel length in MOSFETs was reduced.[7] Although it is not clearly known what influence grain boundaries and subboundaries have on device characteristics, the authors believe that perfect single crystalline SOI films are necessary to increase device process reporducibility and device characteristics stability.[8] From the above points of view, the SOI films recrystallization with a seeding area was carried out. This paper reports the seeding effects for SOI films crystallinity.

2. Experimentals and Discussions

Si films recrystallization on SiO_2 was carried out using a strip heater annealer. Figure 1 shows a diagram of a strip heater annealer, which has a fixed graphite bottom heater with SiC coating for heating substrates up to about 1300°C maximum, and a movable tungsten rod heater. A 2 inch diameter sample is laid on the SiC coated carbon plate, electrically insulated from the bottom heater, so that it can be homogeniously heated. The tungsten rod heater is 2 mm in diameter and 200 mm in length and is set above about 1.5 mm from the sample surface. It can be moved at 0.2 to 2.5 mm/s speeds in parallel to the sample surface.

For investigating the seeding effects for SOI films crystallinity, two kinds of samples were prepared. One sample has variously sized rectangular SOI islands, enclosed by a 5 μm wide stripe seeding area, patterned on (100) Si substrates, as shown in Fig. 2(a). The other sample has a seeding area formed at one side of a (100)Si substrate, as shown in Fig. 2(b). Figure 2(c) shows a cross sectional view of the sample, in which 0.8 μm thick SiO$_2$ was thermally grown by LOCOS method and a 0.5 or 1.0 μm thick poly-Si film was deposited by LPCVD. A 1.0 to 2.0 μm thick SiO$_2$ film was used as a cap film to suppress Si films agglomeration. A bilayer cap, consisting of an Si$_3$N$_4$ layer on an SiO$_2$ layer, was used in preliminary experiments. When the Si$_3$N$_4$ layer was thin, i.e., 30–50 nm, there was no appreciable difference in suppression of Si films agglomeration, compared with SiO$_2$ cap cases. However, if the Si$_3$N$_4$ layer was thicker than 100 nm, Si substrates warped during strip heater recrystallization. Therefore, only an SiO$_2$ film was used as a cap. All samples were pre-annealed at

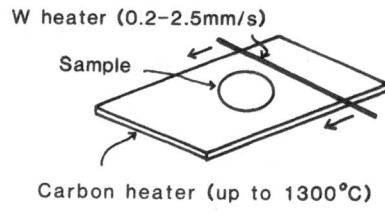

W heater (0.2–2.5mm/s)

Sample

Carbon heater (up to 1300°C)

Fig. 1.   Strip heater annealer diagram.

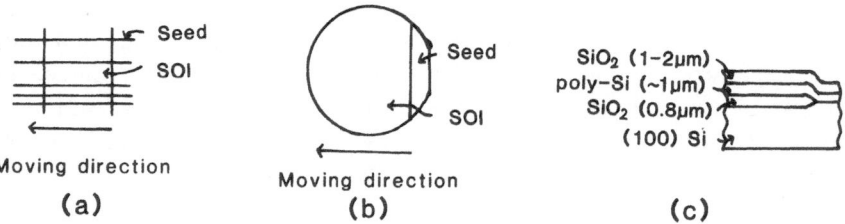

Seed
SOI
Moving direction
(a)

Seed
SOI
Moving direction
(b)

SiO$_2$ (1–2μm)
poly-Si (~1μm)
SiO$_2$ (0.8μm)
(100) Si
(c)

Fig. 2.   Samples structure. (a) SOI islands surrounded by 5 μm stripe seeding. (b) SOI island with seeding area at one side. (c) A cross sectional view.

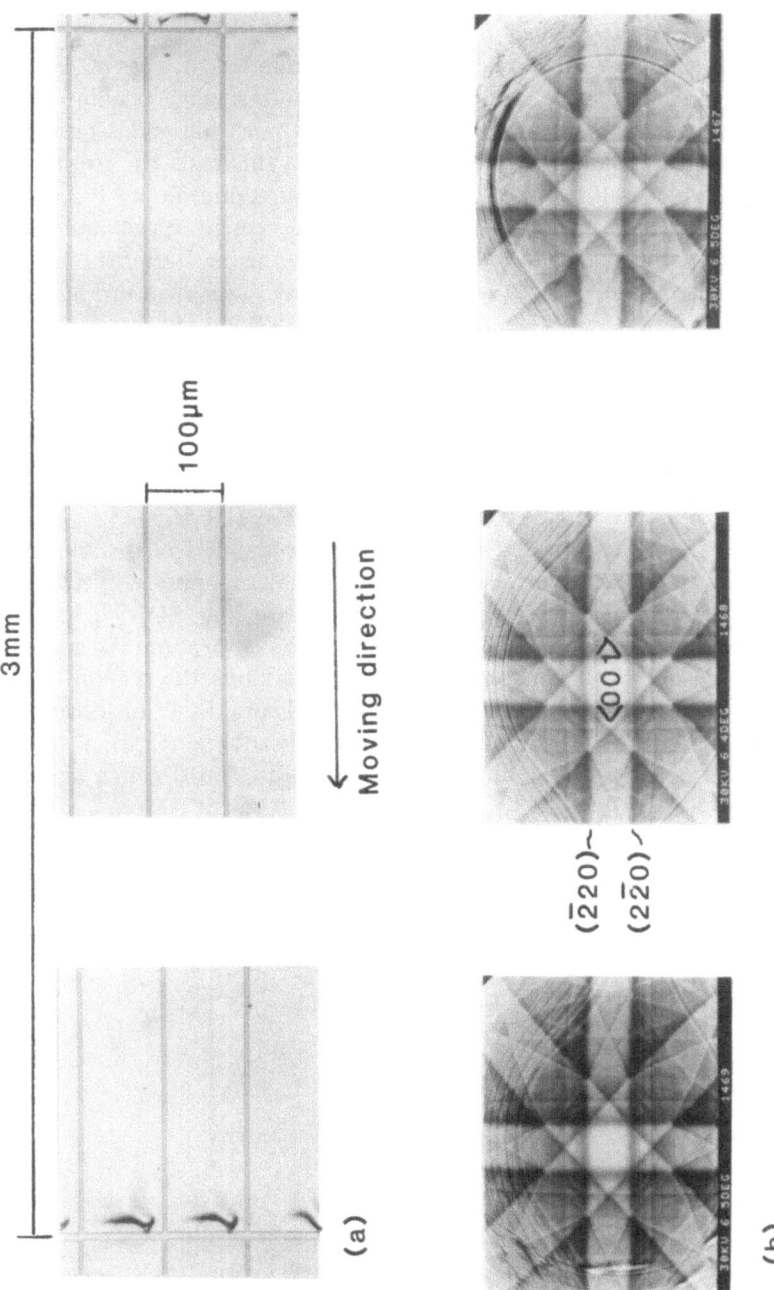

Fig. 3. Optical micrograph. (a) For 100 μm wide and 3 mm long recrystallized SOI islands after Secco etching. (b) ECP at each region in Fig 3(a)

1000°C in N_2 gas for 20 minutes to obtain good adhesion with substrates.

2.1 SOI islands surrounded by a seeding area

For less than 200 μm wide SOI islands surrounded by a stripe seeding area, SOI islands were successfully recrystallized into single crystalline films. Figure 3(a) shows an optical micrograph of 100 μm wide and 3 mm long recrystallized SOI islands after Secco etching. In this case, the top heater moved in parallel to the stripe seeding ($//\langle 110\rangle$direction) at $V = 0.2$ mm/s and $T_s \approx 1200$°C, in which V is top heater movement speed and T_s is substrate temperature. From ECP (Electron Channeling Pattern) observations, SOI islands were found to be single crystals, as shown in Fig. 3(b).

In this manner, SOI islands with less than 200 μm width were easily recrystallized into single crystals. However, in such small SOI islands recrystallization, especially under the slow top heater movement speed condition (about 0.2 mm/s), the substrate surface under SiO_2 films sometimes melted. This surface melting caused small sized SiO_2 films between substrate and SOI films to move. Figure 4 shows a 10 μm wide SiO_2 film sinking, under the condition where $V = 0.2$mm/s and $T_s \approx 1200$°C.

In the case of 0.5 to a few mm wide SOI islands, SOI recrystallization into single crystals became difficult. Figure 5(a) shows an optical micrograph of a part of a recrystallized 1 mm wide and 3 mm long SOI island after Secco etching, under the condition where V is 2.5 mm/s and Ts is about 1200°C. In this case, heater moving direction is along the seeding stripe ($//\langle 100\rangle$). Many subboundaries were grown. From ECP observations, shown in Figs. 5(b)–(d), the orientation of each subgrain normal to the substrate was nearly $\langle 100 \rangle$ direction within 2 or 3° difference and the

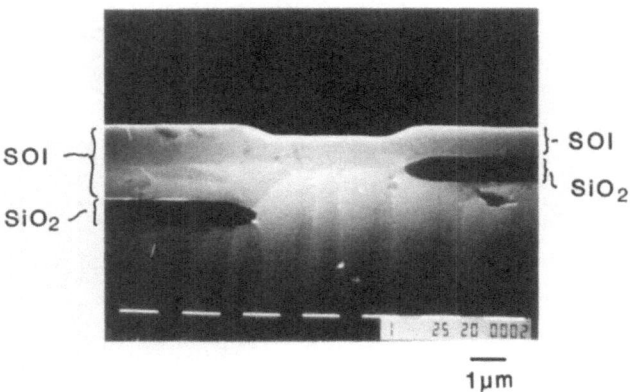

1μm

Fig. 4.   SEM micrograph of cross sectional recrystallized 10 μm wide SOI islands (SiO_2(1.0 μm)/poly-Si(1.0 μm)/SiO_2(0.8 μm)/Si substrate) near a seeding area. On the left, SiO_2 film is sinking about 1μm.

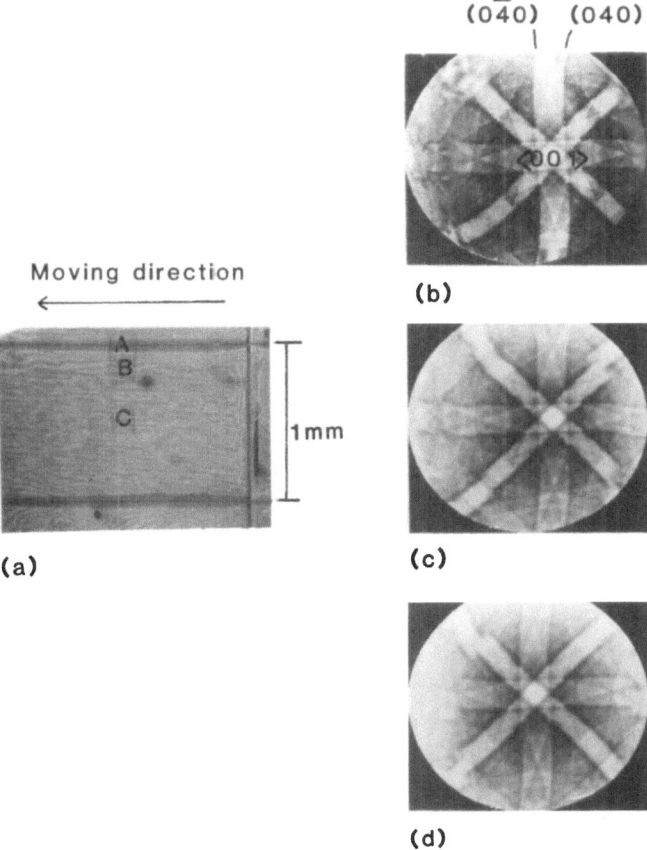

Fig. 5. Optical micrograph (a) for a part of a 1 mm wide and 3 mm long recrystallized SOI island after Secco etching. ECPs at each grain in Fig. 5(a). ECPs (b) for a seeding area at point A, (c) for a adjacent grain at point B and (d) for a grain at point C.

orientation in plane was ⟨100⟩ direction within 0.5°. Grains with large angle difference were not observed.

2.2 SOI islands with seeding area at one side

When a top heater was set in parallel and moved perpendicularly to the seed-SOI boundary, as shown in Fig. 2(b) for one side seeding case, seeded single recrystallization grew laterally over an area only about 100 μm long from the seed-SOI boundary, as shown in region(A) in Fig. 6(a). This laterial epitaxial region stopped at small grains region(B). Recrystallized SOI films (region(C) in Fig. 6(b)), which had dendritic grain boundaries, were observed adjacent to region (B). Figure 6(b) shows an SEM micrograph

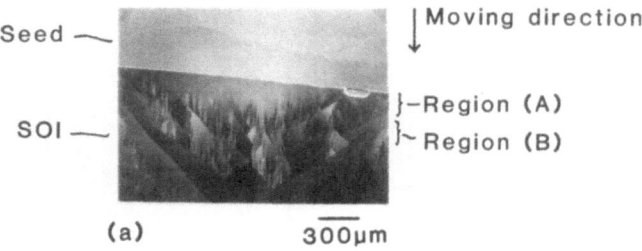

Seed

SOI

Moving direction

}–Region (A)
}–Region (B)

(a)   300μm

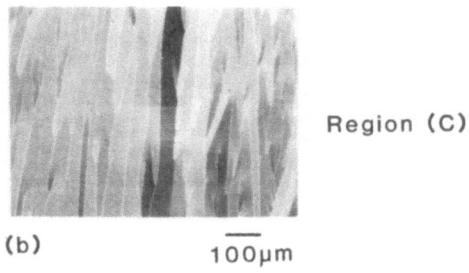

Region (C)

(b)   100μm

Fig. 6. SEM micrographs. (a) Near the seed-SOI boundary. (b) About 5 mm from the seed-SOI boundary.

of region (C) about 5 mm from the seed-SOI boundary. In this region, a few mm wide large angle grains were grown. Each grain contained subgrains. Between these subgrains overall variation in orientation is less than 0.5° in the film plane and less than 5° in the direction normal to the substrate. However, the crystal orientation for a large angle grain distributed randomly.

3. Summary

Recrystallized Si films crystallinity on SiO₂ by strip heater method was found to be dependent on the width of SOI islands surrounded by a seeding area. When (100) substrates were used and heater moving direction was ⟨100⟩ or ⟨110⟩ (in parallel to the stripe seeding), the following results were obtained.

1. SOI islands less than 200 μm wide were recrystallized into single crystalline islands.

2. In 0.5 to a few mm wide SOI islands, subgrains were formed.

3. For one side seeding case, which corresponds to SOI islands with

infinite width, only a 100 μm long single crystal grew laterally from the seed-SOI boundary. Far from the send-SOI boundary, a few mm wide large angle grains, which contained subgrains, were formed.

Acknowledgements

The authors gratefully thank Y. Okuto for his encouragement and helpful discussions and T. Takekawa and K. Yamada for sample fabrications.

This work was performed under the management of the R & D Association for Future Electron Devices as a part of the R & D Project of Basic Technology for Future Industries sponsored by Agency on Industrial Science and Technology, MITI.

REFERENCES

1) J. C. C. Fan, B-Y. Tsaur, R. L. Chapman, and M. W. Geis: Appl. Phys. Lett. **41** (1982) 186.
2) M. W. Geis, H. I. Smith, B-Y. Tsaur, J. C. C. Fan, D. J. Silversmith, and R. W. Mountain: J. Electrochem. Soc. **129** (1982) 2812.
3) R. F. Pinizzotto, H. W. Lam, and B. L. Vaandrager: Appl. Phys. Lett. **40** (1982) 388.
4) B-Y. Tsaur, J. C. C. Fan, and M. W. Geis: Appl. Phys. Lett. **41** (1982) 83.
5) H. W. Lam, R. F. Pinizzotto, S. D. S. Malhi, and B. L. Vaandrager: Appl. Phys. Lett. **41** (1982) 1083.
6) B-Y. Tsaur, J. C. C. Fan, R. L. Chapman, M. W. Geis, D. J. Silversmith, and R. W. Mountain: IEEE Electron Device Lett. **EDL-3** (1982) 398.
7) T. Nishimura, Y. Akasaka, A. Ishizu, and H. Nakata: Appl. Phys. Lett. **42** (1983) 102.
8) J. P. Colinge, E. Demoulin, D. Bensahel, G. Auvert, and H. Morel: IEEE Electron Device Lett. **EDL-4** (1983) 75.

Silicon-on-Insulator: Its Technology and Applications, edited by S. Furukawa, pp. 159–167.
© KTK Scientific Publishers, Tokyo, 1985.

SINGLE CRYSTAL GERMANIUM ISLAND FORMATION ON INSULATOR BY ZONE MELTING

M. Takai, T. Tanigawa, K. Gamo, and S. Namba

Faculty of Engineering Science, Osaka University,
Toyonaka, Osaka 560, Japan

Abstract Single crystal germanium islands on insulator have been formed by zone melting with graphite strip heaters. Two types of geometrical patterns for islands were successfully used to obtain large single crystal islands. One of the patterns was a series of rectangles with 100 μm × 80 μm sides connected with a narrow stripe of 10 μm × 30 μm, which could select and transfer a single grain orientation during melting-zone moving and suppress subboundary formation in the connected island. A second type of an island was an isolated one with a peaked side so that a single crystal could grow from here as a self-seed and melting-zone could move from this side. Both diffraction patterns using TEM and etch-pit patterns after Superoxol etching showed that most islands were single crystal and the predominant orientation was ⟨100⟩. Hall effect measurements on these single crystal germanium islands showed electron concentrations of 10^{16}–10^{18} /cm^3 with Hall mobility of 5×10^3–10^3 cm^2/Vsec.

1. Introduction

Single crystal silicon-on-insulator (SOI) structures by various techniques such as zone-melting with graphite strip heaters, laser and electron beam recrystallization have recently been investigated extensively because of the potential for applications in thin film transistors for flat panel display devices or for realizing three dimensional integrated circuits.[1] Single crystal germanium (Ge) formation on insulator, on the other hand, has not been studied much,[2] though this structure is useful not only for infrared detectors but also for an interface layer of gallium arsenide (GaAs) epitaxy on insulator, in which hybrid integrated circuits including optical functions will be realized. Although recrystallization techniques such as zone-melting using graphite strip heaters or electron beams have been investigated since 1960's,[3–5] it was only recently that these techniques drew attention because of technological needs for application and intensive studies by laser and electron beams. Zone melting recrystallization can provide a larger grain size in comparison with laser recrystallization.[6] However, it is necessary to use well defined geometrical patterns to control grain sizes.[7,8]

In this study, single crystal Ge islands were formed on silicon dioxide by zone-melting using graphite strip heaters. Optical, transmission electron microscopies and Hall effect measurements were performed to characterize isolated Ge islands. Two types of geometrical patterns of islands were successfully used, so that large single crystal islands could be obtained.

2. Experimental Procedures

A 700 nm Ge layer was evaporated from Ge ingots (99.999%) on the thermal oxide of 730 nm thickness of (100) Si, and was mesa-etched by using photolithography to obtain island patterns. About 500 nm silicon dioxide layer was deposited on the Ge layer as a protective layer as shown in Fig. 1. Two types of geometrical patterns for Ge islands were used to obtain large single crystal Ge layers. One of the patterns was a series of rectangles with 100 μm $\times$ 80 μm sides connected with a narrow stripe of 10 μm $\times$ 30 μm, which would select and transfer a single grain orientation during melting-zone moving and suppress subboundary formation in the connected island. A second type of an island was an isolated one with a peaked side so that a single crystal could grow from here as self-seed and melting-zone was moved from this side.

Zone melting was performed in a vacuum chamber ($\sim 10^{-5}$ Torr) with two separate carbon heaters, the temperatures of which were 800°C and 2000–2500°C, respectively. A sample was placed on the movable lower heater with temperature of 800°C and was scanned with a speed of 0.2–1.0 mm / sec. A gap between the rigid upper heater and movable lower heater was 0.5 mm. An experimental setup is schematically shown in Fig. 2.

Optical   microscopy   observation   after   Superoxol (HF:H$_2$O$_2$:H$_2$O = 1:1:10) etching of Ge islands and diffraction pattern measurements using transmission electron microscopy (TEM) were performed to characterize Ge islands. Four points Van-der-Pauw methods

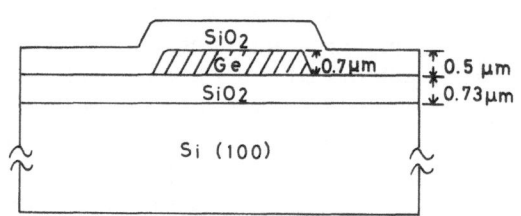

Fig. 1. A cross sectional view through a sample.

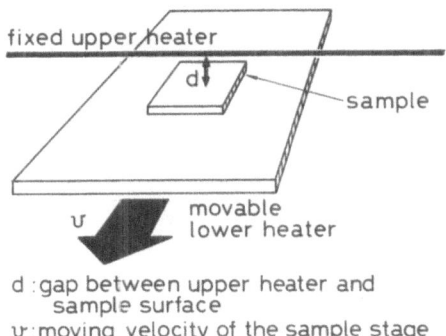

d : gap between upper heater and
    sample surface
v : moving velocity of the sample stage

Fig. 2.   A schematic diagram of zone-melting equipment.

for Hall effect measurements were used to study electrical characteristics of Ge islands. The temperature dependence of carrier density and Hall mobility was also studied.

3.   Results and Discussion

One of the geometrical patterns for Ge islands used was a series of rectangles connected with a narrow stripe as shown in Fig. 3, which showed smooth surfaces after zone melting and Superoxol etching if a connecting stripe was not broken. If this stripe was broken during melting-zone moving, a connected island showed micro subboundaries as in the upper part of Fig. 3-a, indicating the existence of micro grains in the islands. Figure 4 shows micrographs of connecting stripes after zone melting and Superoxol etching. Square-shaped etch-pit patterns indicating ⟨100⟩ orientation were transferred through a narrow stripe, so that the connected island had the same crystal orientation as shown in Fig. 4-a. Triangle-shaped etch-pit patterns indicating ⟨111⟩ orientation, on the other hand, was observed in the connected island if the connecting stripe was broken as in Fig. 4-b. These results indicate that the narrow connecting stripe between rectangle-shaped islands can select and transfer a single grain orientation during melting-zone moving and suppress subboundary formation in the connected island. It was found that a series of single crystalline islands with 100 μm × 80 μm sides could grow with this narrow stripe reproducibly.

A second type of islands used was an isolated island with a peaked side so that a single crystal could grow from here as a self-seed and melting-zone could move from this side. Figure 5 shows optical micrographs of zone-melted islands after Superoxol etching. A single crystalline island was obtained when a peaked side was not evaporated or broken as shown

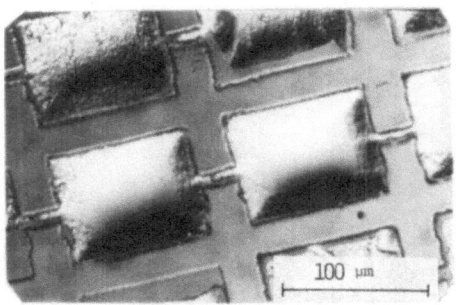

(a)

(b)

Fig. 3.   Optical micrographs of zone-melted islands after Superoxol etching.

in Fig. 5-a, whereas subboundaries were observed when the peaked side was evaporated or broken because of higher heater temperature as in Fig. 5-b. This result indicates that the peaked side is indispensable to single crystalline island formation by zone melting, where it acts as a seed for recrystallization.

Figure 6 shows a histogram for crystal orientations of zone-melted islands obtained from etch-pit pattern observation using Superoxol etchant and corresponding diffraction patterns by TEM. The predominant orientation was found to be ⟨100⟩. In the case of graphoepitaxy of Ge using zone-melting, the predominance of ⟨100⟩ orientation was also observed.[7]

Scan direction

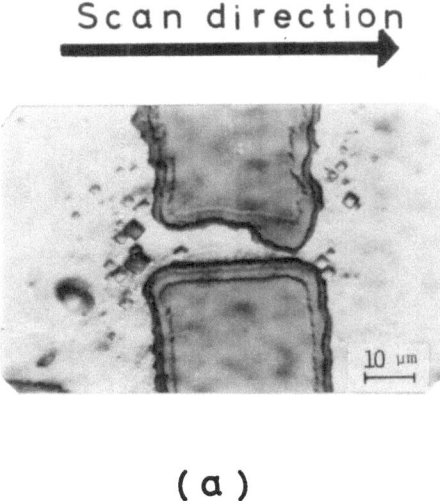

(a)

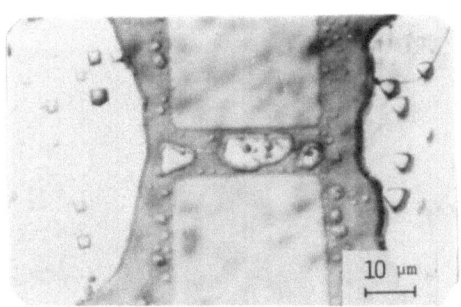

(b)

Fig. 4. Optical micrographs of a connecting stripe between islands after Superoxol etching.

Hall effect measurements on deposited polycrystalline Ge layer show-ed p-type conduction as was already reported.[5] N-type conduction was observed after zone melting, and Hall mobility as a function of carrier density was shown in Fig. 7. The values taken from Irvin[9] are also shown for comparison. The carrier concentration ranged from 10^{16} to 7×10^{17} cm^{-3}. The impurity which gives n-type conduction in this case was not identified. It might be included in Ge ingots and activated during crystalliza-

Scan direction

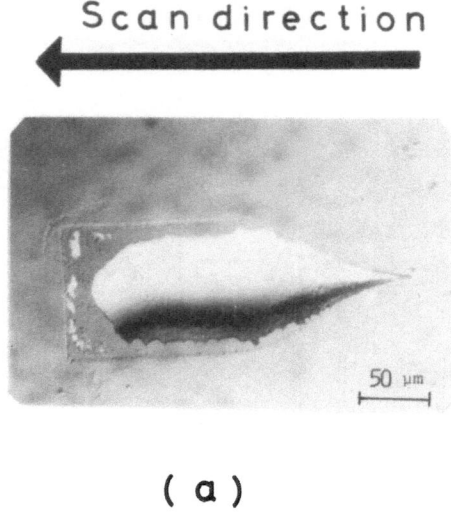

50 μm

(a)

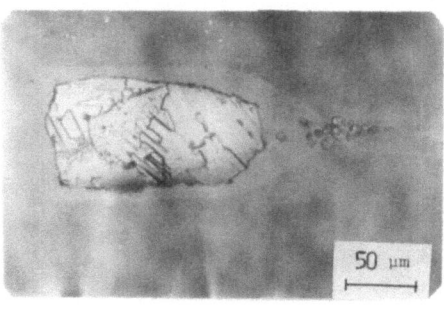

50 μm

(b)

Fig. 5. Optical micrographs of zone-melted islands after Superoxol etching.

tion. The values of Hall mobility were smaller than those obtained in the bulk material by about 50%, which suggested strong compensation. The stress in the Ge film might change those values as found in the case of Si on insulator.[10] The values for islands which have grain boundaries were found to be smaller than those obtained in the bulk material by an order of magnitude.

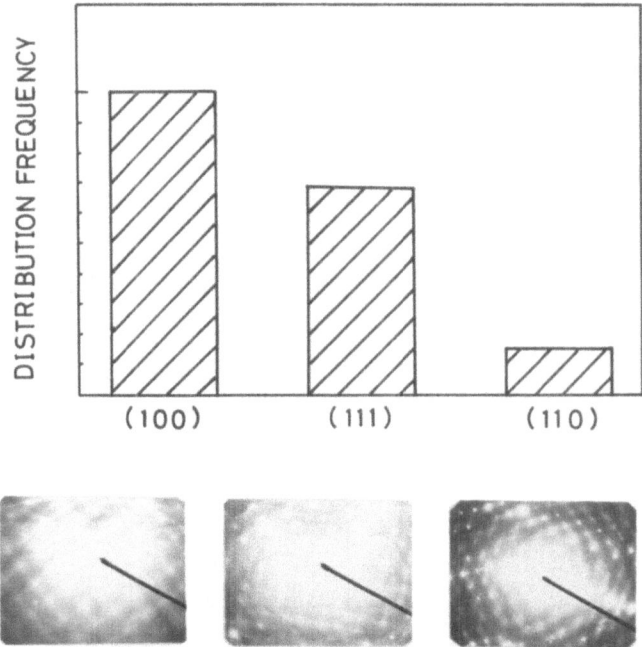

Fig. 6. A histogram for crystal orientations of zone-melted islands and corresponding diffraction patterns by TEM.

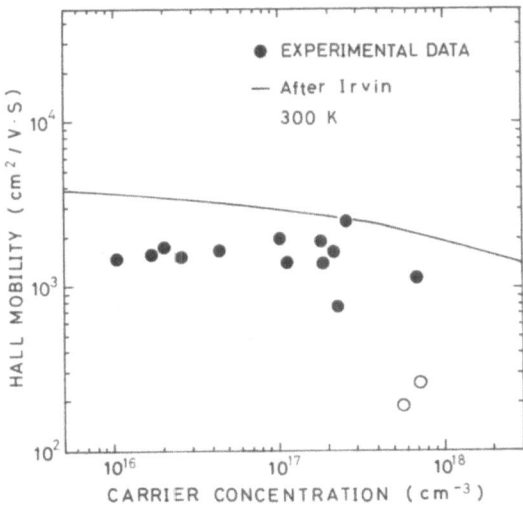

Fig. 7. Hall mobility as a function of carrier concentration for zone-melted Ge. A closed circle was obtained from single crystal islands. An open circle was obtained from islands with boundaries.

Figure 8 shows temperature dependence of Hall mobility and carrier concentration for zone-melted Ge islands. The Hall mobility curve is almost proportional to $T^{-3/2}$ plot from 300 K down to 100 K, but it begins to depart from it at and below 100 K, which indicates the increase in scattering by ionized impurities. The turning point to intrinsic conduction shown in Fig. 8 is higher than that obtained in the bulk material. These results suggest that electrical characteristics of Ge islands are affected by impuri-

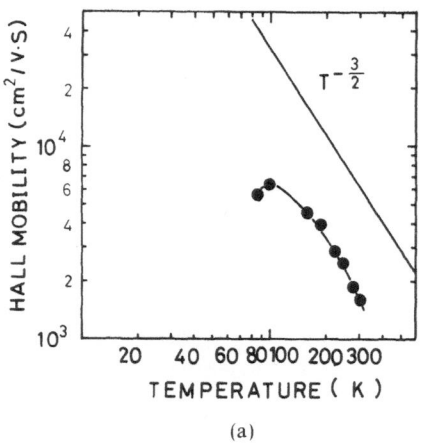

(a)

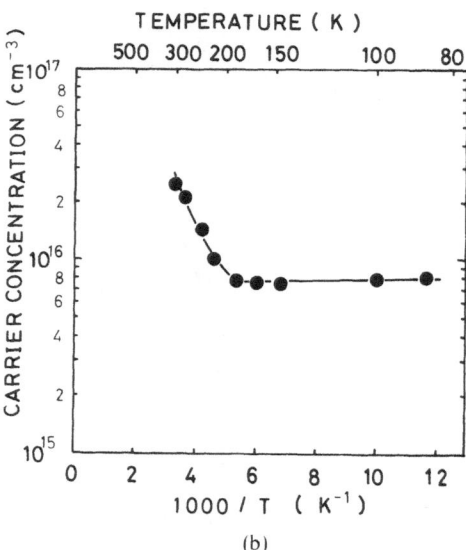

(b)

Fig. 8.   Temperature dependence of Hall mobility and carrier concentration for zone-melted Ge.

ty compensation and the scattering in the carrier density obtained in Fig. 7 is due to this compensation.

4. Conclusion

Single crystal germanium islands on silicon dioxide were formed by zone melting with graphite strip heaters. A series of Ge rectangles with 100 μm × 80 μm sides connected with a narrow stripe or a Ge island with a peaked side could give rise to single crystal islands. Islands patterns used were found to be able to select and transfer a single grain orientation. The predominant concentrations of 10^{16}–10^{18} / cm^3 with Hall mobility of 5×10^3–10^3 cm^2/Vsec were obtained.

Acknowledgements

The authors wish to thank K. Harakawa for his help in designing and assembling zone-melting equipments in the early stage of experiments. The authors are also indebted to Y. Yuba for helpful discussion and to K. Kawasaki for experiments.

REFERENCES

1) See for example, *Laser and Electron-Beam Solid Interactions and Materials Processing*, ed. J. F. Gibbons, L. D. Hess, and T. W. Sigmon (North-Holland, New York, 1981); *Laser and Electron-Beam Interactions with Solids*, ed. B. R. Appleton and G. K. Celler (North-Holland, New York, 1982).
2) J. C. C. Fan, H. J. Ziegler, R. P. Gale, and R. L. Chapman: Appl. Phys. Lett. **36** (1980) 158.
3) G. B. Gilbert, T. O. Poehler, and C. F. Miller: J. Appl. Phys. **32** (1961) 1597.
4) J. Maserjian: Solid-State Electron. **6** (1963) 477.
5) S. Namba: J. Appl. Phys. **37** (1966) 1929.
6) M. W. Geis, H. I. Smith, B-Y. Tsaur, J. C. C. Fan, D. J. Silversmith, and R. W. Mountain: J. Electrochem. Soc. **129** (1982) 2813.
7) K. Sakano, K. Moriwaki, H. Aritome, and S. Namba: Jpn. J. Appl. Phys. **21** (1982) L636.
8) H. A. Atwater, H. I. Smith, and M. W. Geis: Appl. Phys. Lett. **41** (1982) 747.
9) S. M. Sze and J. C. Irvin: Solid-State Electron. **11** (1968) 599.
10) B-Y. Tsaur, J. C. C. Fan, and M. W. Geis: Appl. Phys. Lett. **40** (1982) 322.

CHAPTER 3 : SOLID PHASE EPITAXY

Silicon-on-Insulator: Its Technology and Applications, edited by S. Furukawa, pp. 171–185.
© KTK Scientific Publishers, Tokyo, 1985.

MODELING OF INTERFACE ATOMIC ARRANGEMENT FOR ANALYSIS OF SOLID PHASE EPITAXY AND Si-ON-INSULATOR STRUCTURE

T. SAITO and I. OHDOMARI

*School of Science and Engineering, Waseda University, 3-4-1, Ohkubo,
Shinjuku-ku, Tokyo 160, Japan*

Abstract   We have analyzed an atomic mechanism of solid phase epitaxy (SPE) by an amorphous Si/crystalline Si (a-Si/c-Si) interface model and Si-on-Insulator structure by a c-Si/a-SiO$_2$ interface model. The a-Si/(100)c-Si interface model was constructed by building a continuous random network (CRN) model of a-Si on a (100)c-Si lattice. It consists of 121 atoms on the amorphous side and 230 atoms on the crystalline side. Using the interface model, we have quantitatively analyzed a bond rearrangement process (BRP) at the interface during SPE. We have studied a BRP based on the mechanism by Spaepen and Turnbull (1979). We have simulated nine steps of the BRP. The atomic coordinates at each step of the BRP have been calculated by the energy relaxation based on Keating potential. We have found that the atomic displacements during the BRP are smaller than Si-Si bond length. The BRP results in decrease in distortion energy mainly due to bond bending. The amount of decrease is in good agreement with the value evaluated by the experimental heat of crystallization. The c-Si/a-SiO$_2$ interface model has been constructed by connecting a c-Si lattice and a CRN model of a-SiO$_2$. It contains no dangling bonds and no SiO$_x$ transition region. We have found that the distortion of c-Si lattice is smaller in the c-Si/a-SiO$_2$ model than in the a-Si/(100)c-Si interface model.

1. Introduction

Modeling of atomic arrangement of solid-solid interface is a starting point for understanding structural and electrical properties of the interface, and phase change and reaction at the interface. In this paper, we report two models for interface between crystalline and amorphous solids. They are an amorphous silicon/(100) crystalline silicon [a-Si/(100)c-Si] interface model and a c-Si/a-SiO$_2$ interface model for analysis of solid phase epitaxy (SPE) and Si-on-Insulator (SOI) structure, respectively.

SPE of ion-implanted and deposited a-Si layers on c-Si substrates has been studied by many investigators[1-4] in recent years. The various experimental observations on SPE of a-Si have been examined, such as effects of growth temperature, crystalline orientation and incorporated impurities on SPE rate. However, the atomic mechanism of SPE has not been well understood. In order to interpret the experimental results and improve the technique of SPE, it is necessary to understand the atomic mechanism of SPE of a-Si. Constructing an a-Si/c-Si interface model is a starting point for understanding the atomic mechanism. For this purpose, Spaepen[5] constructed an a-Si/(111)c-Si interface model, and we constructed that of a-Si/(100)c-Si interface.[6] We summarize the procedure of modeling and the characteristics of our model, and quantitatively analyze a bond rearrangement process (BRP), which takes place at the interface during SPE, using the model.

Formation of SOI structure, which is necessary for fabricating three dimensional integrated circuits, by means of lateral SPE of a-Si over SiO_2 film attracts attention recently.[7,8] One of the key factors, which governs the lateral SPE, is a bond formation process between c-Si film and underlying a-SiO_2 film. For understanding the process, it is necessary to examine an atomic structure of c-Si/a-SiO_2 interface in the SOI structure. For this purpose, we have constructed a c-Si/a-SiO_2 interface model. A brief description of the analysis of the interface model is given.

2. Modeling of a-Si/(100)c-Si Interface

2.1 Procedure of modeling

The procedure of modeling an a-Si/(100)c-Si interface is as follows, (i) construct a ball-and-spoke model of interface atomic arrangement; (ii) measure the atomic coordinates of the model; (iii) relax the atomic coordinates to minimize distortion energy by a computer program.

To construct a ball-and-spoke model of the interface, we built a continuous random network (CRN) model[9] of a-Si on a (100) c-Si substrate. The interface model constructed by this method consists of 121 atoms on the amorphous side (121-atom CRN model) and 230 atoms on the crystalline side. It contains no dangling bonds. Figure 1 shows schematically the atomic configuration on the top (100) plane of the c-Si substrate in the a-Si/(100)c-Si interface model. Figures 1-(a) and (b) show a chair-type six-fold ring and a seven-fold ring formed on the (100) plane, respectively. Figure 1-(c) shows a top view of the atomic configuration on the whole (100) plane. Among the atoms above the (100) plane, the atoms on seven-fold rings [e.g. the atoms B and D in Fig. 1-(c)] are on the amorphous side, but the atoms on chair-type six-fold rings [e.g. the atoms A and E in Fig. 1-(c)] are still on the crystalline side. Therefore, the interface

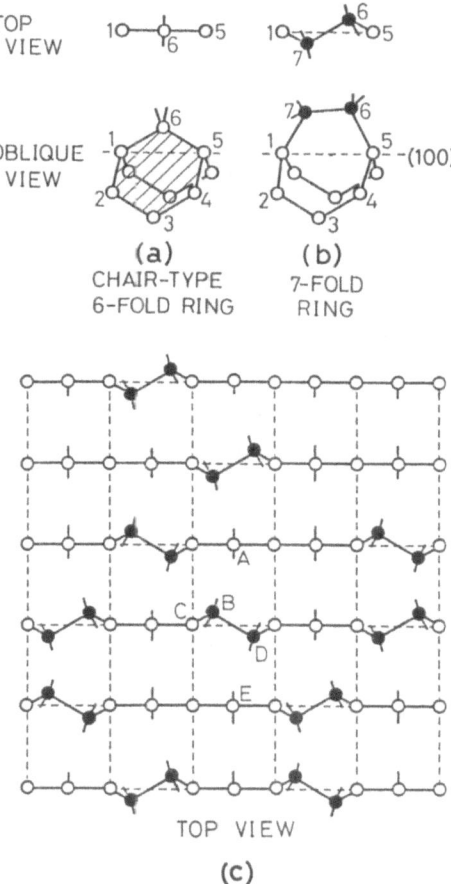

Fig. 1. The atomic configuration on the top (100) plane of c-Si substrate in the a-Si/(100)c-Si interface model. (a) and (b) show a chair-type six-fold ring (shaded area) and a seven-fold ring on the (100) plane, respectively. The dashed line indicates the top (100) plane. The open and the closed circles represent the atoms on the crystalline side and the atoms on the amorphous side, respectively. (c) Top view of the atomic configuration on the whole (100) plane. In these figures, the lower layer atoms in the c-Si lattice are not shown in the top views.

is somewhat rough. In contrast, the a-Si/(111)c-Si interface constructed by Spaepen[5] is sharp, since no atoms are found on the crystalline side above the top (111) plane of c-Si lattice.

The atomic coordinates of the ball-and-spoke model of a-Si/(100) c-Si interface were measured using a ruler. We input them into the computer.

The energy relaxation is accomplished by minimizing Keating[10] expression for the distortion (elastic) energy in terms of the bond lengths and bond angles,

$$
E = \frac{1}{2} \sum_{i[j]} \frac{3\alpha}{8d_0^2} (|\mathbf{x}_i - \mathbf{x}_j|^2 - d_0^2)^2
$$
$$
+ \sum_{i[j,k]} \frac{3\beta}{8d_0^2} \left[(\mathbf{x}_j - \mathbf{x}_i) \cdot (\mathbf{x}_k - \mathbf{x}_i) + \frac{1}{3} d_0^2 \right]^2, \tag{1}
$$

where d_0 is the equilibrium bond length and $\mathbf{x}_i$ are the atomic coordinates. The first sum is over all atoms i and their four neighbours j, the second sum is over all atoms i and all pairs of distinct neighbours j and k. The force constants α and β are related to bond stretching and bond bending, respectively. We used $d_0 = 2.3500$Å, $\alpha = 4.85 \times 10^4$ dyn/cm and $\beta/\alpha = 0.2$ in the relaxation. In the relaxation process, each atom is moved in turn along the direction of the force on it, a distance proportional to that force. The force on atom i is given by

$$
\mathbf{F}_i = -\nabla_i E. \tag{2}
$$

After each movement, the forces are recalculated, and the atoms moved again, and the process is iterated until the displacements become small. The bond-length and bond-angle deviations are simultaneously relaxed by this procedure. Figure 2 shows the atomic arrangement of the a-Si/(100)c-Si interface model obtained by the energy relaxation program.

2.2 Characteristics of the model

Using the atomic coordinates obtained by the energy relaxation program, we examined the characteristics of the a-Si/(100)c-Si interface model.

In the 121-atom CRN model on the (100) c-Si lattice, a r.m.s. bond-length deviation of 1.05% and a r.m.s. bond-angle deviation of 6.86° were found. For the relaxed 519-atom CRN model by Steinhardt et al.,[11] a r.m.s. bond-length deviation of 1.04% and a r.m.s. bond-angle deviation of 7.10° were reported. The 121-atom CRN model has the same magnitude of distortion as the 519-atom model. We also found that the bond-angle distribution of the 121-atom model is roughly symmetrical about the perfect tetrahedral angle and that the dihedral angle distribution is continuous from the eclipsed (0°) to the staggered (60°) indicating a slight preference for the staggered configuration. The calculated reduced intensity function $F(s)$ of the model is in good agreement with the experimental result. From these results, we concluded that the 121-atom CRN model has characteristics very similar to those of the bulk CRN model.

We also examined the bond-angle deviations in the c-Si lattice and on the boundary between the c-Si and the CRN model. The principal

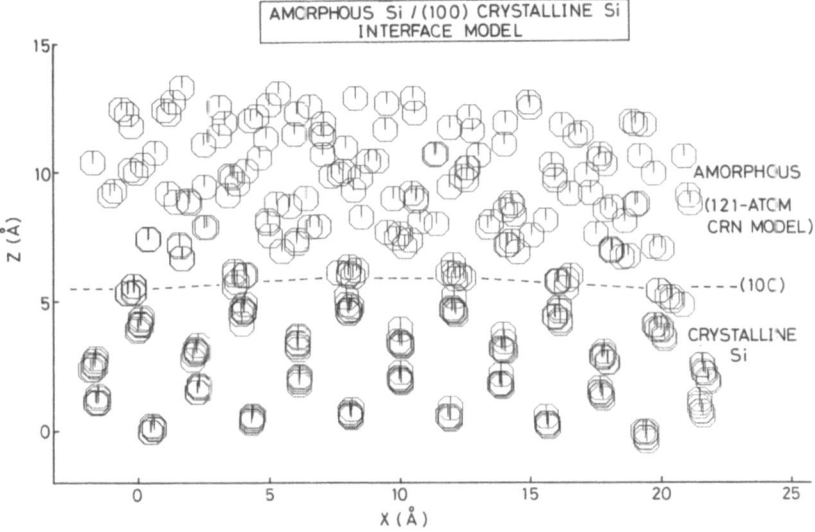

Fig. 2. The a-Si/(100)c-Si interface model. The atomic arrangement was obtained by the energy relaxation program.

results characterizing the interface are (a) the r.m.s. bond-angle deviation in the c-Si is 3.52^c (not zero), and (b) the r.m.s. bond-angle deviation on the boundary is $9.05°$ indicating a larger distortion than in the bulk CRN model. The details of the procedure of the modeling and characteristics of the model described hitherto are given in the ref. 6.

Recently, we have examined the distortion energy distribution (DED)[12] in the interface model for better understanding of distortion at the interface. We have calculated a bond bending DED over atoms, which is a distribution of the values of bond bending distortion energy of an atom (E_b) for all four-coordinated atoms in the model as proposed by Grigorovici.[12] A value of E_b is calculated by summing up the values of bond bending distortion energy of an angle belonging to that atom, using the second term of eq. (1). Figure 3 shows the results of the calculation, where DED's in the 121-atom CRN model and in each layer of (100) c-Si lattice are shown in the separate histograms. In the 121-atom CRN model, the DED roughly fit $f_5(\chi^2)$ function, which means that the CRN model is truely random. This is the same result as that of the precise analysis of relaxed Polk-Boudreaux model[13] examined by us.[14] In the c-Si lattice, both the mean and the maximum values of E_b are largest in the first layer. Moreover, the mean value is larger in the first layer than in the a-Si. (The variation of the mean value in the model is somewhat different from Fig. 13 in ref. 6, since we have used only four-coordinated

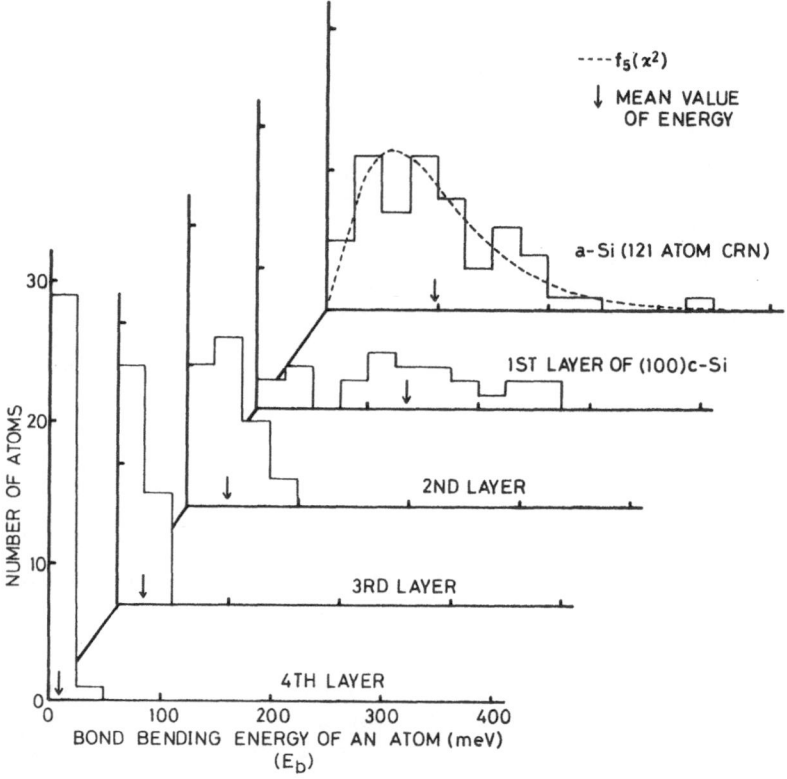

Fig. 3. The bond bending DED over atoms in the a-Si/(100)c-Si interface model. The adapted $f_5(x^2)$ function, which is the chi-square distribution function with the degree of freedom 5, is also shown.

atoms in the case of the DED calculation.) The large distortion at the interface implies high frequency of bond breaking at the interface during SPE. Using the interface model, we have quantitatively analyzed a BRP, which transforms a CRN structure of a-Si to a diamond cubic structure of c-Si at the interface, in the later sections.

2.3 Bond rearrangement process at a-Si/(100)c-Si interface

We have found a BRP, which proceeds by the Spaepen and Turn-bull's mechanism,[15] at a-Si/(100)c-Si interface using our ball-and-spoke interface model. The BRP simulates the SPE process along ⟨100⟩ direction which has the highest growth rate.[1] In this section, we have described the change in connectivity of bonds during the BRP. The analysis of the BRP is only qualitatively as that of Spaepen and Turnbull.[15] In the Section 2.4, we quantitatively analyze the BRP.

The atomic configuration in the interface model, already shown in Fig. 1 and Fig. 2, is an initial state before the BRP starts, which we call step (0) of the BRP. Figure 4 shows a central part of the model, where the atoms A to G exist, in detail. Since the BRP proceeds in the central region as we show later, it is not affected by the surface of the model.

The BRP starts with breaking a bond between the atoms G and E, and generating two dangling bonds at the interface, as proposed by Spaepen and Turnbull.[15] This is step (1) of the BRP in Fig. 5-(1). After step (1),

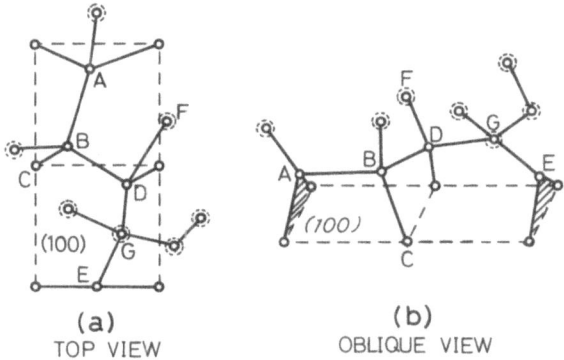

(a)
TOP VIEW

(b)
OBLIQUE VIEW

Fig. 4.   Top and oblique view of a central part of the atomic configuration in Fig. 1-(c). The shaded area represents a part of a chair-type six-fold ring, characteristic of c-Si lattice. The atoms surrounded by the dashed circle are not shown in Fig. 1. The atoms A and E are already on the lattice site, because they are on the chair-type six-fold ring.

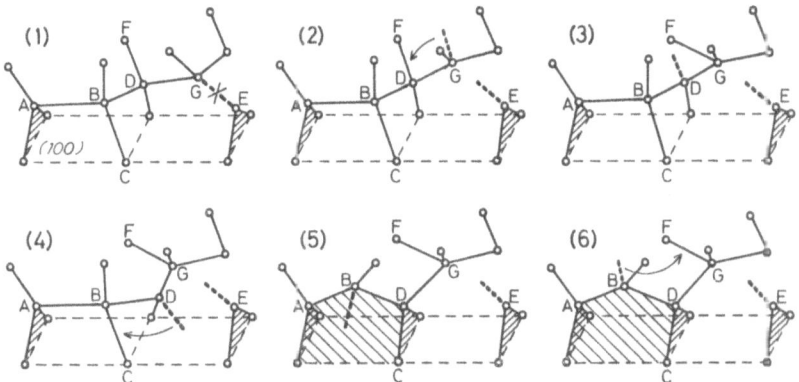

Fig. 5

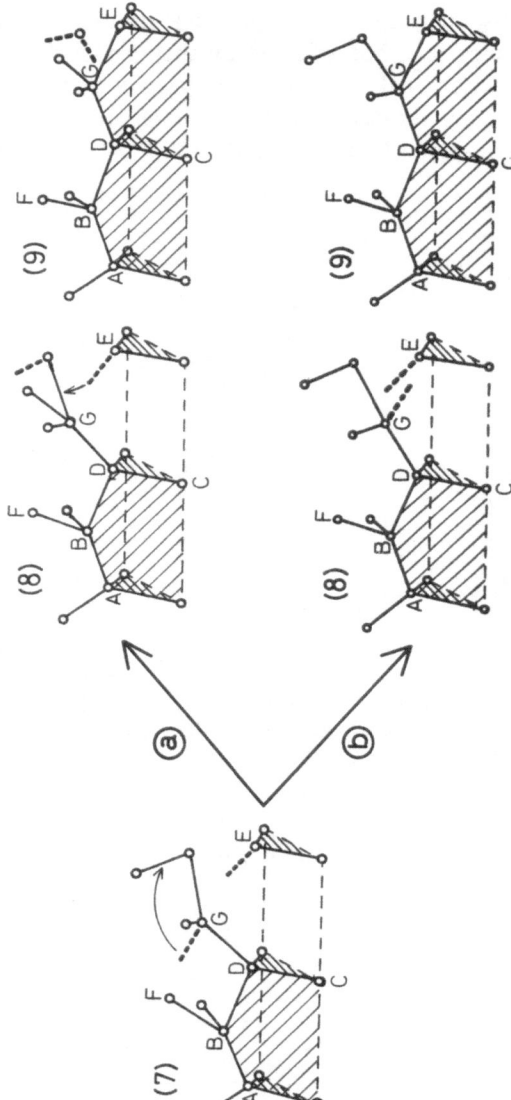

Fig. 5. The bond rearrangement process (BRP) at the a-Si/(100)c-Si interface during solid phase epitaxy. The number in the parentheses denotes the step of the BRP. The shaded area represents a part of a chair-type six-fold ring.

the BRP proceeds by the process of simultaneous bond-breaking and new-bond-formation, as shown in Fig. 5. The dangling bonds move around at the interface. At step (5), new chair-type six-fold rings are formed and the atoms B and D move to the lattice sites. After the step (7), both processes ⓐ and ⓑ are possible. At step (9) on both processes, new chair-type six-fold rings are again formed and the atom G moves to the lattice site. Moreover, at step ⓑ-(9), two dangling bonds are annihilated by the recombination with each other. This is the first demonstration of the recombination process.[16]

Up to the last step of the BRP, three atoms move to the lattice sites in total. Thus the SPE process of a-Si on a (100) c-Si substrate can be explained qualitatively by the BRP. However, at this stage of the study, the amount of decrease in distortion energy, which is expected to occur by the crystallization, is not calculated. It is also unknown whether the amount of temporal increase in distortion energy, which may occur at some of the steps of the BRP, is small enough to be overcome by the thermal activation. Therefore we have carried out a quantitative analysis of the BRP.

2.4 Quantitative analysis of the bond rearrangement process

The method of the quantitative analysis of the BRP is as follows. Firstly, we have calculated the equilibrium atomic coordinates at each step of the BRP based on diagrams in Fig. 5. To calculate the atomic coordinates, we have used the energy relaxation program based on the Keating potential in eq. (1). The distortion energy at the step (n) of the BRP, denoted by $E^{(n)}$, is calculated from the $x_i^{(n)}$'s, where $x_i^{(n)}$ is the atomic coordinates of the i-th atom at step (n). The procedure of the relaxation is similar to that used to relax our interface model described in Section 1. However, since the two dangling bonds exist inside the interface model during the BRP, we have adopted the method of calculation which was used by Marklund[17] to relax structures with dangling bonds. In the method, dangling bonds are saturated by pseudo-atoms. Thus an atom, which has dangling bonds, is considered to have four nearest neighbour atoms in the calculation. Secondly, using the atomic coordinates obtained by the energy relaxation, we have analyzed the BRP. We have calculated atomic displacements and change in the distortion energy during the BRP. An atomic displacement of i-th atom from its position at step $(n-1)$ to that of step (n) is expressed as ; $d_i^{(n)} = |x_i^{(n)} - x_i^{(n-1)}|$. A change in the distortion energy at step (n) from that of step (0) is expressed as; $(\Delta E)^{(n)} = E^{(n)} - E^{(0)}$.

By the energy relaxation, the atomic coordinates at each step of the BRP have converged to equilibrium positions under the interactions of the interatomic forces. For example, Figure 6 shows the projection on

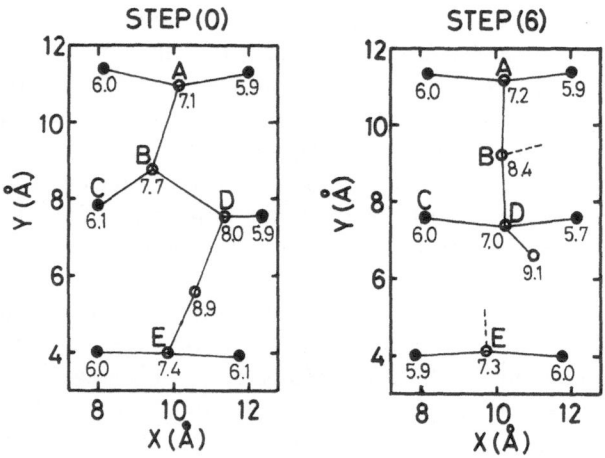

Fig. 6. The projection of the atomic positions at step (0) and (6) of the bond rearrangement process. They are obtained by the energy relaxation program. The closed circles represent the (100) surface layer atoms. The value near the circle represents z-coordinate. The dashed bars indicate dangling bonds.

a (100) plane, of the some of the atomic positions at step (0) and (6).

Using the coordinates, we have calculated the atomic displacement $d_i^{(n)}$ during the BRP. Figure 7 shows the displacement of atoms B and D which move to the lattice site during the BRP. The displacements are always smaller than half of the Si–Si bond length in c-Si. This means that the atomic displacements during the BRP are smaller than that of the self-diffusion process by the vacancy mechanism in c-Si. The atomic displacements are small during the BRP, in spite of the fact that the drastic change in structure take place at the interface due to the crystallization.

Figure 8 shows the change in the distortion energy $(\Delta E)^{(n)}$ during the BRP. Three features are seen in the figure; (i) the distortion energy temporally increases about 1eV for the first six steps of the BRP; (ii) after passing through these steps by the thermal activation, the distortion energy decreases in the process ⓑ; (iii) on the contrary, the distortion energy again increases in the process ⓐ. The last feature indicates quantitatively that process ⓑ is more probable than process ⓐ.

An activation energy required for the BRP, E_{BRP}, which can be compared with the experimentally observed activation energy of SPE, is roughly estimated as follows. E_{BRP} is a sum of two terms; $E_{BRP} = E_{BRP}^1 + E_{BRP}^2$, where E_{BRP}^1 is an energy required for breaking a bond at the step (1), and E_{BRP}^2 is a temporal increase in the distortion energy during the BRP. E_{BRP}^1 is estimated 1.8eV[18] as Spaepen did.[5] The estimate of E_{BRP}^2 is not so easy, because it consists of not ony the temporal increase in the

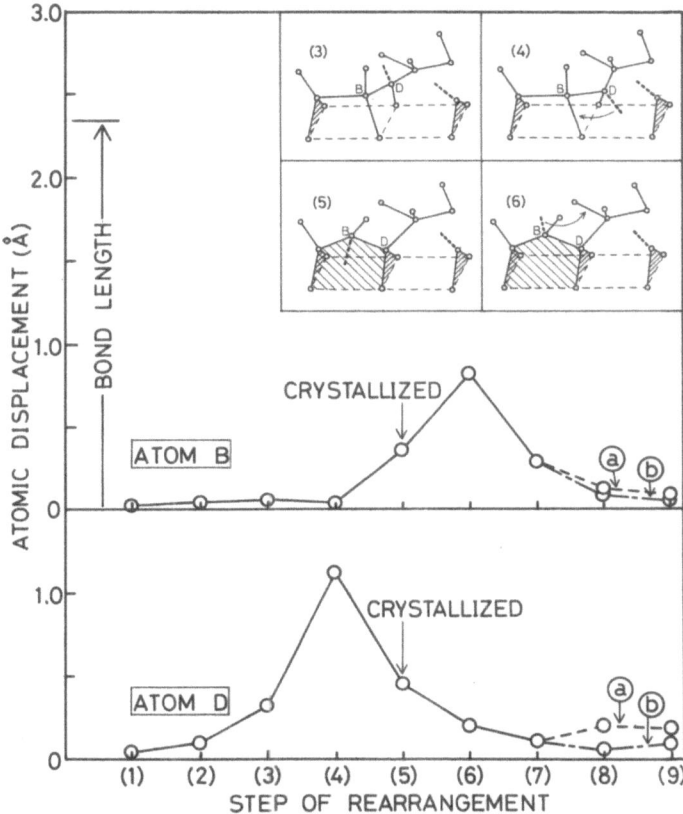

Fig. 7. The atomic displacement $d_i^{(n)}$ for the atoms B and D during the bond rearrangement process.

distortion energy at the first six steps of the BRP but also the temporal increase at the intermediate configuration between the successive steps of the BRP. The former contribution to E_{BRP}^2 is estimated about leV as described in feature (i). However, this value is not very accurate because the distortion energy expressd by Keating potential becomes inaccurate when the distortion is large. For better accuracy, the third order term in Keating potential[19] is needed. The latter contribution to E_{BRP}^2 is considered to be fairly lower than 1.8eV which is the energy required for breaking a bond. This is because the energy increase at the intermediate configuration is equal to that for the process of simultaneous bond-breaking and new-bond-formation, which is postulated a low energy process.[16] Therefore, E_{BRP}^2 is a mid value between 1.0 and 2.8eV. Consequently, E_{BRP} is roughly estimated at a value between 2.8 and 4.6eV. However,

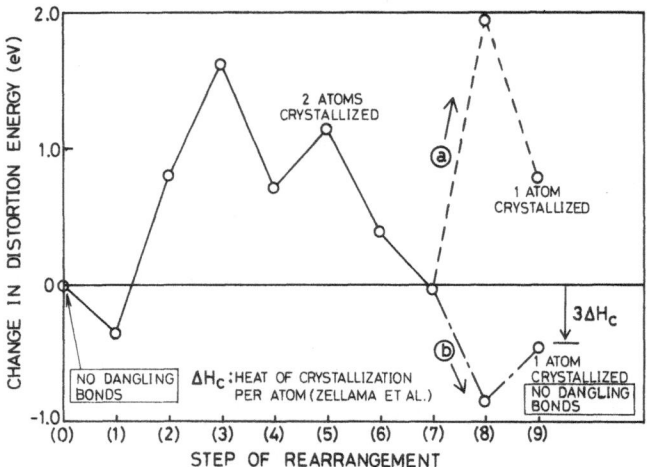

Fig. 8.   The change in the distortion energy $(\Delta E)^{(n)}$ during the bond rearrangement process.

the estimated value of E_{BRP} is right order of magnitude and within factor two of the experimental activation energy of $2.3 \sim 2.85$ eV.[1,4]

At step ⓑ-(9), the change in the distortion energy takes a negative value due to the energy decrease by crystallization. Up to this step, three atoms have moved to the lattice sites in total. Therefore, the amount of decrease in the distortion energy is expected to be three times the heat of crystallization per atom (ΔH_c). $3\Delta H_c$ evaluated by the experimental result of Zellama et al.[20] is denoted by a short bar in the figure. As is clearly seen in the figure, the amount of the decrease in the distortion energy is in good agreement with the experimentally obtained value, $3\Delta H_c$. Moreover, we have found that the decrease in the distortion energy comes almostly from the decrease in the bond-bending energy. The fact that the energy released during the crystallization is almost due to bond bending is consistent with the results of the calculation on the distortion energy of relaxed Polk-Boudreaux[13] model of a-Si by us.[14] It is concluded quantitatively that the decrease in energy due to crystallization takes place at the end of the BRP.

3.   Modeling of c-Si/a-SiO₂ Interface

The procedure of modeling c-Si/a-SiO$_2$ interface is similar to that of the a-Si/(100)c-Si interface model. We have constructed a ball-and-spoke model of c-Si/a-SiO$_2$ interface by connecting a (100) c-Si lattice and a CRN model of a-SiO$_2$.[21] The size of the model is nearly the same as that of the a-Si/(100)c-Si interface model. It contains no dangling bonds and no SiO$_x$ transition region. The effects caused by the difference in the thermal expansion coefficients between c-Si and a-SiO$_2$ are not taken into account. Therefore, the model simulates an ideal SOI structure. The concept of the model is schematically shown in Fig. 9. We have measured the atomic coordinates and relaxed them by a computer program using Keating type potential. The form of potential function is similar to that used by Ching[22] to relax periodic models of a-SiO$_2$, and SiO$_x$, which includes the bond stretching constant α and the bond bending constants of Si and oxygen atoms, β_1 and β_2, respectively.

Using the atomic coordinates obtained by the energy relaxation, we have calculated the variations of distortion energy due to bond-stretching and bond-bending per Si and oxygen atoms as a function of distance from the interface. We have found that the bond stretching energy is negligibly small at the interface. In contrast, the bond bending energies of both Si and oxygen atom increase at the interface. The energy is largest on a plane where c-Si and a-SiO$_2$ meet. However, the distortion of c-Si lattice, which is mainly due to bond bending, is smaller in the c-Si/a-SiO$_2$ model than in the a-Si/(100)c-Si interface model. This is probably due to the flexibility of bond angles of oxygen atoms in the c-Si/a-SiO$_2$ model.

For understanding the mechanism of the lateral SPE over SiO$_2$ film, we can analyze a bond formation process of c-Si/a-SiO$_2$ interface based on the model, as well as a BRP at a laterally moving a-Si/c-Si interface. The details of the characteristics of the model and its application to the analysis of the lateral SPE will be published elsewhere.

4. Conclusion

We have constructed the a-Si/(100)c-Si interface model and the c-Si/a-SiO$_2$ interface model for analysis of SPE and SOI structure, respectively. The ball-and-spoke models were constructed and their atomic coor-

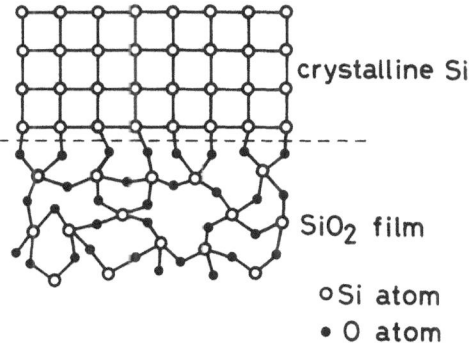

crystalline Si

SiO$_2$ film

○ Si atom
● O atom

Fig. 9.   The schematic view of c-Si/a-SiO$_2$ interface model. It simulates an ideal Si-on-Insulator structure.

dinates were relaxed by a computer program using Keating potential.

In the a-Si/(100)c-Si interface model, the 121-atom CRN model has similar structural properties to the bulk CRN model, concerning the bond length and bond angle deviations, bond angle and dihedral angle distributions and $F(s)$. The characteristic excess distortion at the interface is found. Using the interface model, we have carried out a quantitative analysis of the BRP at the interface. We have examined nine steps of the BRP, which proceeds by the mechanism suggested by Spaepen and Turnbull.[15] The atomic coordinates during the BRP have been obtained by energy relaxation. Using the coordinates, we have calculated the atomic displacements during the BRP. They are always smaller than Si–Si bond length. The BRP results in decrease in the distortion energy by crystallization. The amount of decrease is in good agreement with the value evaluated by the experimental heat of crystallization. We conclude quantitatively that the BRP studied in this paper explains the process of SPE of a-Si on (100) c-Si substrate.

In the c-Si/a-SiO$_2$ interface, we have found that the bond bending distortion energy of Si and oxygen atoms concentrates at the interface, while the bond stretching energy is negligibly small. The application of the model to the analysis of the SOI process, such as the lateral SPE, is now in progress.

REFERENCES

1) S. S. Lau, J. W. Mayer, and W. Tseng: *Handbook on Semiconductors, Vol. 3,* ed. S. P. Keller (North-Holland, 1980) p. 531.
2) I. Ohdomari, M. Kakumu, H. Sugahara, M. Hori, T. Saito, T. Yonehara, and Y. Hajimoto: J. Appl. Phys. **52** (1981) 6617.
3) S. A. Kokorowski, G. L. Olson, and L.D. Hess: J. Appl. Phys. **53** (1982) 921.
4) A. Lietoila, A. Wakita, T. W. Sigmon, and J. F. Gibbons: J. Appl. Phys. **53** (1982) 4399.
5) F. Spaepen: Acta Metall. **26** (1978) 1167.
6) T. Saito and I. Ohdomari: Philos. Mag. **B43** (1981) 673.
7) H. Yamamoto, H. Ishiwara, S. Furukawa, M. Tamura, and T. Tokuyama: *Poc. 14th Symp. on Ion Implantation and Submicron Fabrication* (held at Institute of Physical and Chemical Research, 1983) p. 73.
8) Y. Kunii, M. Tabe, and K. Kajiyama: J. Appl. Phys. **54** (1983) 2847.
9) D. E. Polk: J. Non-Cryst. Solids **5** (1971) 365.
10) P. N. Keating: Phys. Rev. **145** (1966) 637.
11) P. Steinhardt, R. Alben, and D. Weaire: J. Non-Cryst. Solids **15** (1974) 199.
12) R. Grigorovici: J. Non-Cryst. Solids **35–36** (1980) 1167.
13) D. E. Polk and D. S. Boudreaux: Phys. Rev. Lett. **31** (1973) 92.
14) T. Saito, T. Karasawa, and I. Ohdomari: J. Non-Cryst. Solids **50** (1982) 271.
15) F. Spaepen and D. Turnbull: *Laser-Solid Interactions and Laser Processng,* ed. S. D. Ferris, H. J. Leamy, and J. M. Poate (American Institute of Physics, New York, 1979) p. 73.
16) V. J. Fratello, J. F. Hays, F. Spaepen, and D. Turnbull: J. Appl. Phys. **51** (1980) 6160.
17) S. Marklund: Phys. Status Solidi (b) **100** (1980) 77.

18) L. Pauling: *Nature of the Chemical Bond* (Cornell Univ. Press, Ithaca, 1960) p. 85.
19) P. N. Keating: Phys. Rev. **149** (1966) 674.
20) K. Zellama, P. Germain, S. Squelard, J. C. Bourgoin, and P. A. Thomas: J. Appl. Phys. **50** (1979) 6995.
21) R. J. Bell and P. Dean: Philos. Mag. **25** (1972) 1381.
22) W. Y. Ching: Phys. Rev. **B26** (1982) 6610.

Silicon-on-Insulator: Its Technology and Applications, edited by S. Furukawa, pp. 187–207.
© KTK Scientific Publishers, Tokyo, 1985.

LATERAL SOLID PHASE EPITAXY OF EVAPORATED AMORPHOUS Si FILMS ONTO SiO$_2$ PATTERNS

H. Yamamoto,[1] H. Ishiwara,[1] S. Furukawa,[1] M. Tamura,[2]
and T. Tokuyama[2]

[1]*Department of Applied Electronics, Tokyo Institute of Technology,
Nagatsuda, Midoriku, Yokohama 227, Japan*
[2]*Central Research Laboratory, Hitachi Ltd., Kokubunji, Tokyo 185,
Japan*

Abstract Lateral solid phase epitaxy (L-SPE) of amorphous-Si (a-Si) films vacuum-evaporated on Si substrates with SiO$_2$ patterns has been investigated, in which the film first grows vertically in the regions directly contacted to the Si substrates and then grows laterally onto SiO$_2$ patterns. It has been found from transmission electron microscopy and Nomarski optical microscopy that use of dense a-Si films, which are formed by evaporation on heated substrates and subsequent amorphization by Si$^+$ ion implatation, is essentially important for L-SPE. The maximum L-SPE length of 5–6 μm was obtained along the $\langle 010 \rangle$ direction after 10hour-annealing at 600°C. The kinetics of the L-SPE growth has also been investigated.

1. Introduction

Single or large-grained recrystallization of silicon films on insulating substrates (SOI) has been successfully realized by liquid phase techniques such as laser,[1] electron beam,[2] or carbon heater annealing.[3] These methods, however, may not be applied to fabrication of 3-dimensional LSI's, since the surface of the recrystallized layer is usually not smooth and since the impurity profile in the underlying Si layer is probably deformed by thermal diffusion during the heating process to melt the top Si film. An alternative approach to realize the SOI structure at low temperature is lateral solid phase epitaxy (L-SPE) of a deposited amorphous Si (a-Si) film, in which the film first grows vertically in the regions directly contacted to the Si substrate and then grows laterally onto SiO$_2$ patterns, until the growth is stopped by random crystallization as schematically shown in Fig. 1.

Ohmura *et al.*[4] first reported L-SPE of vacuum-evaporated a-Si films

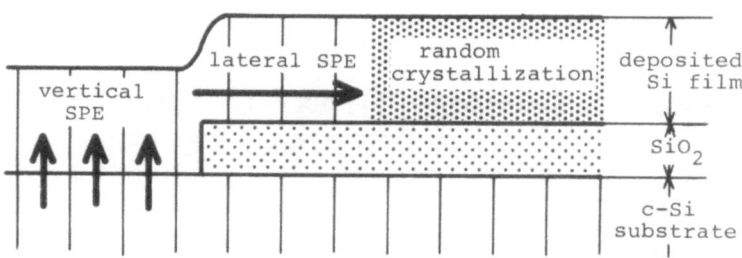

Fig. 1.   Schematic diagram showing formation of an SOI structure by lateral solid phase epitaxy.

onto SiO_2 patterns. In their case, however, only isolated grown layers of about 1μm wide and 0.5μm long were obtained in the films heavily implanted by Si^+ ions. The small growth area may be due to their experimental conditions that SiO_2 patterns were formed only in the $\langle 011 \rangle$ direction on (100)Si wafers and that the evaporation was not performed in the ultra-high-vacuum (UHV) condition. Kunii *et al.*[5] succeeded to grow "clean-CVD-deposited" a-Si films laterally onto SiO_2 patterns. They used stripe patterns formed parallel to the both $\langle 001 \rangle$ and $\langle 011 \rangle$ directions on the (100)Si wafers and the deposited films were crystallized epitaxially over a 4μm wide $\langle 001 \rangle$ stripe after either 550°C for 9days or 650°C for 1hour annealing. The lateral growth rate observed for the $\langle 010 \rangle$ stripe was somewhat slower than the (100) vetical growth rate, but it was extremely slow for the $\langle 011 \rangle$ stripe.

We have also been studying the L-SPE of UHV-evaporated a-Si films,[6] and have found that the lateral growth onto the SiO_2 patterns formed parallel to the both $\langle 010 \rangle$ and $\langle 01\bar{1} \rangle$ directions on the (100)Si wafers occurs after annealing at 600°C for several hours when dense a-Si films formed by evaporation on heated substrate and subsequent amorphization by Si^+ ion implantation are used. In this paper, we'll review our experimental results and discuss about the necessary conditions and the growth kinetics of vacuum evaporated a-Si films onto SiO_2 patterns.

2.   Estimate of the Growth Length

As can be seen from Fig. 1, the L-SPE is a competitive process between SPE growth and random crystallization. In these two competitors, SPE growth of self-implanted or UHV-evaporated a-Si films on single crystal Si substrates (vertical SPE growth) has been studied extensively by Rutherford backscattering spectroscopy (RBS),[7,8] time-resolved-reflectivity measurement (TRR),[9] and some other techniques. It has been revealed that the SPE growth is thermally activated with the activation

energy of 2.4–2.9eV, and that the growth rate exhibits a strong orientation dependence; the rate on the (100) substrate is 2 to 4 times and 20
to 25 times higher than those on the (110) and (111) substrates, respectively. Some of the reported SPE rates on these three substrates at 600°C
are shown in Table 1. Note that two different rates were reported for
the growth on the (111) substrate.[7] The growth rate of UHV-evaporated
a-Si films on the (100) substrates was measured to be about a half of
that of self-implanted a-Si films.[9,10]

On the other hand, there are only a few reports studied about the
kinetics of the random crystallization, which is composed of the random
nucleation and the growth processes, of UHV-evaporated a-Si films. Blum
and Feldman[11] determined the crystalline fraction of the films evaporated
in high-vacuum and annealed in N_2 atmosphere using the optical absorption measurement and determined a characteristic time of the crystallization (a time at which the crystallization of the film is nearly completed).
Zellama et al.[12] determined the crystalline fraction in the films evaporated
in UHV and exposed to air by the electrical conductance measurement
and derived the random nucleation and growth rates. And Roth et al.[13]
measured a random crystallization time (a time to reach $1 - 1/e$ volume
fraction of crystallites) of the UHV evaporated a-Si films by TRR method.
All these reports show that the random crystallization is thermally activated with the activation energies shown in Table 2. Also shown in the
table are the random crystallization times at 600°C reported by these
authors. The values of the random crystallization time (the time to reach
$1 - 1/e$ volume fraction of crystallites) and its activation energy tabulated
in the second group were determined from the data presented in ref.12.

From this table we can see the following facts.

1) The measured activation energies of random crystallization are
3.1 to 3.8eV and they are higher than those of SPE growth (2.4 to 2.9eV).
It suggests that the lower annealing temperature is preferable to obtain

Table 1. Reported vertical SPE growth rates (r) at 600°C and activation energies (E_a) of
the a-Si films on the (100), (110), and (111)Si substrates.

Sample preparation	Substrate orientation	Temperature range (°C)	E_a (eV)	v (600°C) (cm/s)	Reference
Self impla.	(100)	550–700		8.2×10^{-8}	
	(110)	500–575	2.35	3.3×10^{-8}	7
	(111)	550–575		1.5×10^{-8}	
	(111)	550–600		5.5×10^{-9}	
Self impla. UHV evaporation	(100)	550–950	2.68 2.71	1.03×10^{-7} 5.5×10^{-8}	9

H. Yamamoto *et al.*

Table 2. Reported random crystallization times (t) at 600°C and activation energies (E_a) of the evaporated a-Si films on SiO_2.

T_s(°C)	Deposition rate (nm/s)	Thickness (nm)	Temperature range (°C)	Annealing ambient	E_a (eV)	t (600°C) (hours)	Reference
200 300	0.3–0.5	500	560–700	N_2	3.1	26	11)
RT 300	3	100–200	580–650 560–600	vacuum	3.5	19 3.3	12)
RT	0.1	100	660–900 680–1100	vacuum(*in situ*) air	3.16 3.74	0.73 90	13)

longer L-SPE growth because the random crystallization rate becomes relatively lower at lower temperature.

2) Random crystallization is strongly suppressed by exposure of the samples to air, so that the L-SPE samples should be annealed after being exposed to air if the contamination from air dose not seriously affect the SPE growth rate.

3) The experimental values of the random crystallization time varies very widely due to unknown factors, since the random crystallization mechanism has not yet made clear.

Here, we estimate the L-SPE length in a-Si films under the assumptions that the lateral SPE growth rate is the same as the (100) vertical SPE rate and that the random crystallization is not affected by the presence of the SiO_2 patterns. From Table 1 and 2, it seems reasonable to assume that the growth rate and the random crystallization time at 600°C are larger than 5.5×10^{-8}cm/s and 3.3 hours, respectively, when the samples are exposed to air. These values give an estimated L-SPE length of about 6μm even in the worst case, which is considered long enough to perform primary experiments. Moreover, a somewhat longer L-SPE length was expected when the samples were annealed at lower temperature.

3. L-SPE in a-Si Films Formed by the Conventional Evaporation

3.1 Experimental procedure

Amorphous Si films in the L-SPE experiment were first formed by the conventional UHV-evaporation as usually done in the vertical SPE experiments. In this chapter, we review the results of such experiments.

Si(100) wafers were thermally oxidized and patterns of stripes, square-islands, square-windows and so on were formed by the conventional photolithography process. Thickness of the SiO_2 films were 100–300nm. The pattern sizes ranged from 2 to 20μm and their edges were directed parallel to the $\langle01\bar{1}\rangle$ and $\langle010\rangle$ axes of the Si substrates. In these sample configurations, relatively high L-SPE growth rates on the (110) and (010) planes are expected, because these planes are parallel to the pattern edges and normal to the sample surfaces, as shown in Fig. 2. Particularly, the patterns parallel to the $\langle01\bar{1}\rangle$ axis were mainly used in the first experiments. The wafers were chemically cleaned by HNO_3 acid and RCA solutions and etched slightly by diluted HF (HF:H_2O = 1:19) acid. Reduction of the SiO_2 thickness is estimated to be less than 20nm after the chemical cleaning process. They were then loaded in a vacuum chamber equipped with ion pumps and heated at 800°C for 30min to flash clean the bare Si regions. On these cleaned wafers, a-Si films of 300–500nm thick were deposited at room temperature by e-gun-evaporation of polycrystalline Si source with 0.1ppm purity. The pressure during evaporation was less than 5×10^{-6}

(a)

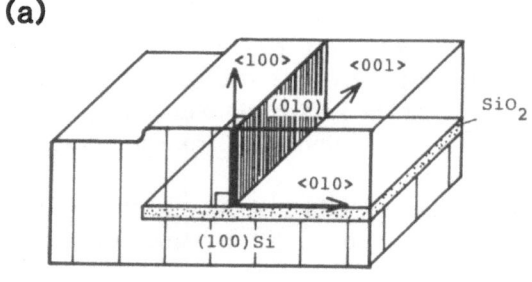

(b)

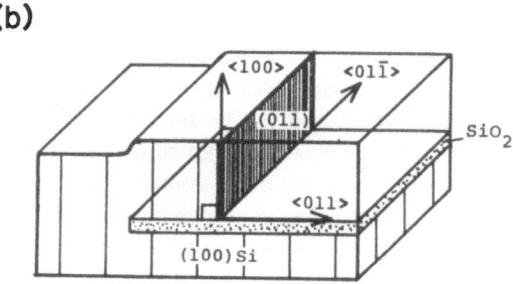

Fig. 2.   Two sample configurations. (a) SiO₂ films on the (100)Si substrates were patterned
parallel to the 〈001〉 axis. (b) SiO₂ films on the (100)Si substates were patterned parallel
to the 〈011〉 axis. The (010) and (011) planes are parallel to the pattern edges and normal
to the sample surfaces.

Pa and the deposition rate was 0.1 to 0.3nm/s. The deposited samples
were then annealed in the same vacuum at about 400°C for 60min to
decrease the void densities in the films so that the films would not absorb
ambient gases when exposed to air.[14] The samples were finally annealed
using an electric furnace in N₂ atmosphere at temperatures ranging from
550 to 650°C for up to 20 hours to induce vertical (V-) and L-SPE. The
annealing temperature around 600°C was mainly used, since, as discussed
previously, the higher temperature is not preferable to L-SPE, while the
lower temperature needs extremely long annealing period. The grown
samples were investigated by RBS, transmission electron microscopy (TEM),
scanning electron microscopy (SEM), and Nomarski optical microscopy.

3.2   Results

First, samples with a deposited-Si(400nm)/patterned-SiO₂
(100nm)/(100)Si structure were annealed at 600°C for 12hours. A typical
TEM image is shown in Fig. 3. We can see from this figure that the
a-Si film grew almost perfectly by V-SPE in the bare Si regions but the

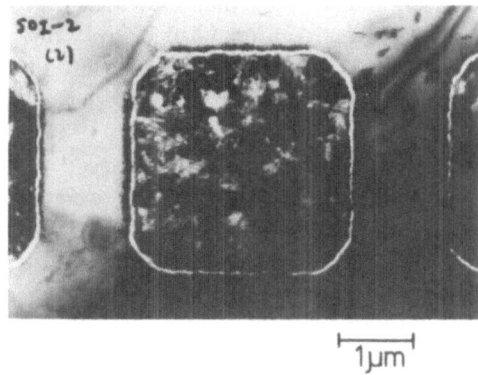

Fig. 3.   TEM image (top view) for a sample of a-Si(400nm)/patterned-SiO$_2$(100nm)/(100)Si structure annealed at 600°C for 12hours. The inner parts of $4 \times 4\mu$m squre patterns whose edges are parallel to the $\langle 011 \rangle$ and $\langle 01\bar{1} \rangle$ axes are the Si/SiO$_2$ regions.

growth was completely stopped at the edge of the underlying SiO$_2$ patterns and the film was polycrystallized in the Si/SiO$_2$ regions. That is, no lateral growth was observed in these samples, although the V-SPE was almost perfect. Better results might be obtained if the patterns parallel to the $\langle 001 \rangle$ axis were used, since the L-SPE on the patterns parallel to the $\langle 011 \rangle$ axis was reported to be difficult.[5] However, we couldn't observe any L-SPE even in the circle patterns. So, there should be other factors to prevent L-SPE in these samples.

We, then, annealed Si(520nm)/patterned-SiO$_2$(100nm)/(100)Si samples at 600°C and measured the times required to finish the V-SPE in the Si/Si regions and to produce the random crystallization in the Si/SiO$_2$ regions. The V-SPE was examined by optical microscopy since the crystallization is detectable by the change of the color. It was also examined by RBS. The result for the V-SPE time was about 15min, which means that the V-SPE rate in the Si/Si regions is about 6×10^{-8} cm/s or somewhat higher when the heat-up time is taken into account. This rate is equal to or somewhat higher than the reported one for UHV-evaporated a-Si films on flat (without SiO$_2$ patterns) (100)Si substrates.[9] From this result, we can say that the V-SPE is not considerably affected by the presence of the SiO$_2$ patterns.

On the other hand, the random crystallization was examined by Nomarski optical microscopy after the grain boundaries were revealed by Wright-etching. The resultant crystallization time was about 60min. Comparing these SPE and random crystallization times, we can see that there is enough time to grow a length of about 1μm by L-SPE if it starts immediately after the completion of the V-SPE and if its rate is identical to that of the V-SPE on the (110) plane. This fact indicates that the SPE growth was stopped at the pattern edge regions, i.e., it is very difficult to crystallize the a-Si films in these regions. In addition, the measured random crystallization time was far shorter than the reported ones shown in Table 2 except a value measured for the *in situ* annealed films.[13] This result seems to show that the random crystallization was enhanced by the presence of the SiO_2 patterns. In order to check this effect, we investigated the random crystallization in the films formed on the flat-SiO_2/Si structures.

The reflectivity of the Si/SiO_2/(100)Si samples was measured by He–Ne laser light (normal incidence) as a function of isothermal annealing time in N_2. Our method to determine the crystalline fraction was essentially the same as the TRR method,[13] though the samples were annealed in a conventional electric furnace and the reflectivity was not measured *in situ*. Figure 4 shows a typical result for a sample annealed at 625°C. In this sample, a 290nm thick a-Si film was deposited at room temperature and annealed *in situ* at 400°C, which was the same process as that of the samples for the L-SPE experiments. Thickness of the underlying SiO_2 film, which was made chemically in the H_2O_2-HCl solution, was so thin ($\simeq$ 1nm) that had no effect on the optical reflectivity. We can see from this figure that the reflectivity normalized with the value of c-Si wafers ($\bar{R} = R/R_{c\text{-}Si}$) is higher than unity when the annealing period is shorter

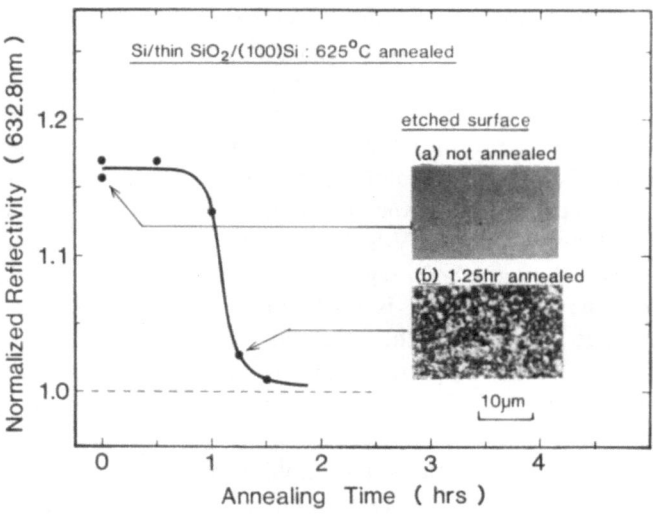

Fig. 4. Random crystallization of the room-temperature-evaporated a-Si films on the flat-SiO_2/(100)Si substrate examined by the reflectivity measurement. The reflectivity was measured by He–Ne laser light (normal incidence) and normalized with the value of c-Si wafers. It was also examined by Wright-etching technique.

than 1 hour and it drops to about unity when the period is longer than
1.5 hours. These two reflectivities are thought to be the ones of an
evaporated a-Si film ($\bar{R}_{\text{a-Si}}$) and a fully polycrystallized film ($\bar{R}_{\text{poly-Si}} = 1$).
And the volume fraction of the poly-crystallites (x) is roughly given by
the next equation.[13]

$$x = (\bar{R}_{\text{a-Si}} - \bar{R})/(\bar{R}_{\text{a-Si}} - 1). \tag{1}$$

Thus, the crystallization time (the time to satisfy $x = 1 - 1/e$) can be determined as $\simeq 1.2$ hours. The crystallization was also examined by the etching
technique. And the result, which is shown in Fig. 4, is essentially identical
to that obtained by the reflectivity measurement. We also checked the
dependence of the crystallization time on the SiO$_2$ thickness. However,
no strong dependence was observed. From a series of similar measurements
at the annealing temperatures of 625 to 675°C, we estimated the random
crystallization time in the a-Si films evaporated at room temperature and
in situ heated at ~ 400°C as ~ 6 hours at 600°C. The activation energy
was also estimated as 3.5 eV. These values are nearly identical to those
for the film evaporated at 300°C.[12] We can see from this result that the
random crystallization time of the films on the patterned-SiO$_2$(100 nm) /
(100)Si structures is about 6 times shorter than that of the films on the
flat SiO$_2$/(100)Si structure. Thus, it has been proved that the random
crystallization is strongly enhanced by the presence of the SiO$_2$ patterns.

3.3 Discussions

So far, we have shown two experimental results on the L-SPE of
room-temperature-evaporated a-Si films on the patterned-SiO$_2$/Si structures, i.e., it is very difficult to crystallize the a-Si films at the SiO$_2$ pattern edge regions and the random crystallization in the films is enhanced
by the presence of the SiO$_2$ patterns. Here, from the cross-section SEM
observation, we show another phenomenon which is thought important
to investigate the factors to prevent L-SPE, that is, the phenomenon that
the Si film of the annealed sample was extremely and rather neatly etched
at the pattern edge regions by Wright etchant as shown in Fig. 5. This
phenomenon can be explained by the enhanced etching due to porosity
of the a-Si films and/or internal stress induced by the porosity. That is,
the a-Si films evaporated at room temperature are thought to be porous
especially at the pattern edge regions because the substrate surface is step-
ped at these regions. Moreover, the porous films probably shrink during
the annealing process in the vacuum chamber (~ 400°C) and/or in the
electric furnace (~ 600°C), and the stress will be produced in the film,
especially at the pattern edge regions. In the case of the sample shown
in Fig. 5, we can presume that the stress produced by the shrinkage during
the SPE growth in the Si/Si regions accumulated to the pattern edge regions

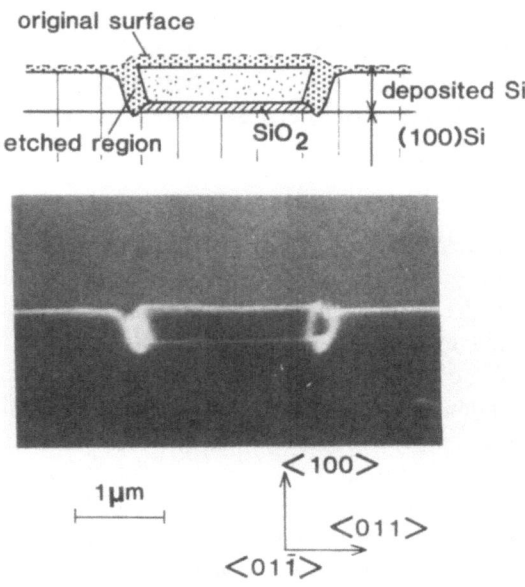

Fig. 5.   Cross-section SEM micrograph for a sample annealed at 600°C for 2hours, cleaved, and etched by Wright etchant. The thicknesses of SiO_2 and evaporated a-Si films are 100nm and 500nm, respectively. The Si film was evaporated at room temperature.

and the film was etched neatly at these regions. It can be thought that such porosity and/or stress are the origin of the above two phenomena and they are essential factors to prevent L-SPE. That is, a stressed film should have a higher random nucleation rate and the high porosity and/or the strong stress at the pattern edge regions should prevent crystallization of the film.

Thus, we presumed that the main factors to prevent L-SPE in the samples with conventional evaporated a-Si films were the porosity of the film and the stress induced by the porosity. If this presumption is correct, some kinds of dense a-Si films are needed for L-SPE. Actually, L-SPE was observed in CVD a-Si films, which are thought to have high density even at the pattern edge regions.[5] However, in order to use CVD a-Si films for L-SPE, the "clean-CVD" apparatus and the precise control of deposition conditions are needed. So, we try to form dense a-Si films by UHV-evaporation and subsequent Si^+ ion implantation.

4.   L-SPE in Dense a-Si Films

4.1   Formation of dense a-Si films

Si films evaporated on heated c-Si and SiO_2 substrates grow in single-

and poly-crystals, respectively. These crystallized films should have such high densities as to be nearly equal to that of bulk c-Si. And the a-Si films formed by subsequent amorphization by Si^+ ion implantation should also have high densities. In this chapter, we optimize the detailed conditions to form dense a-Si films by such processes.

First, to optimize the substrate temperature during evaporation (T_s), 200–300nm thick Si films were evaporated at various T_s on the thin SiO_2 films formed chemically in $H_2O_2 + HCl$ solution and studied by the reflectivity measurement of He–Ne laser light. Figure 6 shows the reflectivity of these films normalized with the value of c-Si wafers. We can see from this figure that the normalized reflectivity of the films is higher than unity when $T_s \lesssim 400°C$ and it drops to about unity at $T_s \simeq 500°C$. This result indicates that the film evaporated at $T_s \simeq 500°C$ is polycrystalline film with fairly large grains. The grain size was actually measured by TEM observation for the films evaporated at $T_s \simeq 500°C$ on 100nm thick SiO_2 films and it was determined to be several tens of nano-meters as shown in Fig. 6. At $T_s \simeq 600°C$, however, the sample surface was roughened and $\bar{R}$ became

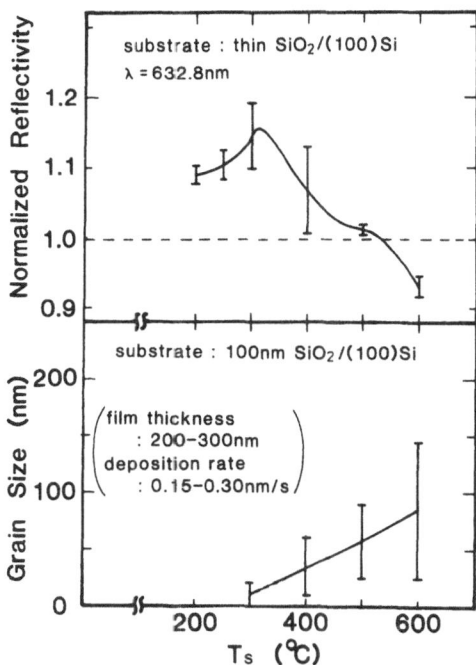

Fig. 6.   Normalized reflectivity and grain size of the Si films evaporated on the $SiO_2/(100)Si$ substrates as a function of the substrate temperature during evaporation. The reflectivity was measured by He–Ne laser light and normalized with the value of the c-Si wafers, while the grain size was measured by TEM observation.

lower than unity. We can say from these results that the optimum substrate temperature is about 500°C.

Next, a 240nm thick film evaporated at $T_s \simeq 500°C$ on the thin-SiO$_2$/(100)Si substrate was implanted with ^{28}Si$^+$ ions to a dose of 2×10^{15} cm^{-2} at both 160 and 60keV. Though the intentional cooling was not employed, the substrate temperature during implantation was considered to be kept near room temperature for the dose rate was very low ($< 0.3\mu A/cm^2$). So, these dose and energies are thought to be sufficient to amorphize the whole evaporated film. This presumption was checked by RBS; the implanted film was studied by 1.5MeV He$^+$ RBS and it was found that the aligned yield was equal to the random yield from the surface to a depth of $\simeq 280$nm. The amorphousness was also checked by the measurement of the random crystallization time. The dependence of the normalized reflectivity on annealing time measured for the implanted film at 625°C is shown in Fig. 7, in which the result in Fig. 4 for the film evaporated at room temperature and *in situ* heated at ~400°C was also shown for comparison. We can see from this figure that the random crystallization time of the implanted film is almost identical to that of the room-temperature-evaporated a-Si films. This result indicates that the implanted film contains negligible amount of micro crystallites and that these implantation conditions are sufficient to amorphize the Si films evaporated at $T_s \simeq 500°C$.

4.2 Experimental procedure for L-SPE

We again tried L-SPE using dense a-Si films formed by the above

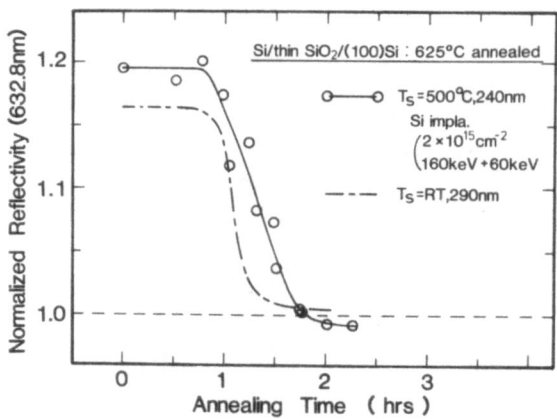

Fig. 7. Normalized reflectivity of an a-Si film evaporated at $T_s \simeq 500°C$ on the thin-SiO$_2$/(100)Si substrate and amorphized by the Si$^+$ ion implantation as a function of the annealing time at 625°C. It is compared with that of a room-temperature-evaporated a-Si film.

process. In addition to that, thinner SiO_2 films of about 50nm thick (the oxidation thickness) were mainly used. The SiO_2 film thickness would be reduced to $\simeq 40$nm after the chemical cleaning process. Both the porosity of the Si film at the pattern edge regions and the stress in the film are expected to be decreased by the reduction of the SiO_2 thickness. The SiO_2 patterns were mainly formed parallel to the $\langle 001 \rangle$ axis of the substrates. The patterns parallel to the $\langle 01\bar{1} \rangle$ axis were also used. Si films of 200–300nm thick were evaporated at $\simeq 500°$C on the patterned wafers and then amorphized by $^{28}Si^+$ ion implantation. The ion energies were varied with the thicknesses of the deposited films, i.e., they were 70 and 180keV for the 250–300nm thick films and 60 and 160keV for the 200–250nm thick films. The dose at each energy was fixed to 2×10^{15} cm^{-2}. These samples are classified as samples A. We also prepared the samples B, in which the a-Si films were deposited at room temperature and heated in the same vacuum at about 400°C. The samples B are compared with the samples described in the previous chapter to check the effect of the reduction of the SiO_2 thickness, as well as being compared with the samples A.

4.3 Results

In Fig. 8 we show a TEM image and TED patterns for a sample A annealed at 600°C for 8hours. In this sample, the $\simeq 40$nm thick SiO_2 patterns parallel to the $\langle 001 \rangle$ axis were used. We can see from this figure that L-SPE occurred in the Si film on SiO_2 near the pattern edge and

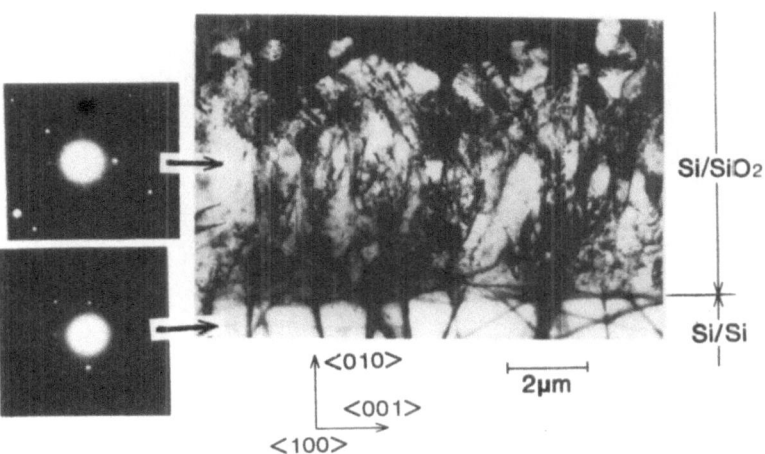

Si/SiO₂

Si/Si

$\uparrow \langle 010 \rangle$

$\langle 001 \rangle \rightarrow$

$\langle 100 \rangle$

2μm

Fig. 8. TEM image (top view) and TED patterns for a sample A annealed at 600°C for 8hours. The SiO_2 stripe patterns are parallel to the $\langle 001 \rangle$ axis and the thickness of the SiO_2 film is about 40nm.

the orientation of the crystallized region is identical to that of the Si substrate. The average L-SPE length along the ⟨010⟩ direction normal to the pattern edge was about 5μm, though the growth front is not smooth. The crystalline quality of the laterally grown area is not so good. The typical defects in the L-SPE region are high-density dislocations and twins.[15] We also show in Fig. 9, a TEM image and TED patterns for the sample A with the patterns parallel to the ⟨011⟩ direction which was annealed at 600°C for 7hours. The SiO₂ thickness in this sample is identical to that of the sample shown in Fig. 8. We can see from this figure that L-SPE also occurs on the patterns parallel to the ⟨01$\bar{1}$⟩ direction. The L-SPE length along the ⟨011⟩ direction was about 3μm and it was shorter than that on the ⟨001⟩ directed patterns. However, there is no clear difference of the roughness of the growth front and the crystalline quality of the grown area between the two samples with different growth directions.

The L-SPE can be also observed by optical microscopy using Wright-etching technique since the etching rate in the ungrown a-Si areas is higher than that in laterally grown areas and the growth fronts of the L-SPE are delineated due to the difference of the etching rates. In Fig. 10, we show Nomarski optical micrographs of the samples A and B annealed at 600°C for 2 to 8hours and etched by Wright etchant. In these samples, the patterns were formed parallel to the ⟨001⟩ direction using the ≃40nm thick SiO₂ films. As can be seen in this figure, the Si films in the samples

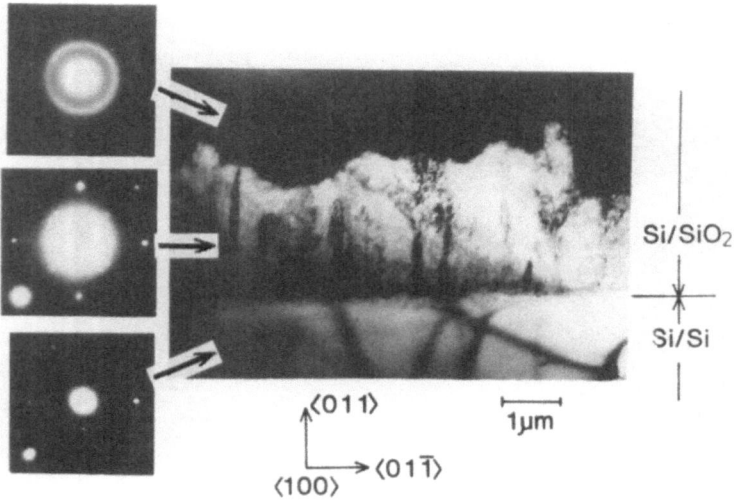

Fig. 9. TEM image (top view) and TED patterns for a sample A annealed at 600°C for 7hours. The SiO₂ stripe patterns are paralell to the ⟨01$\bar{1}$⟩ axis and the thickness of the SiO₂ film is about 40nm.

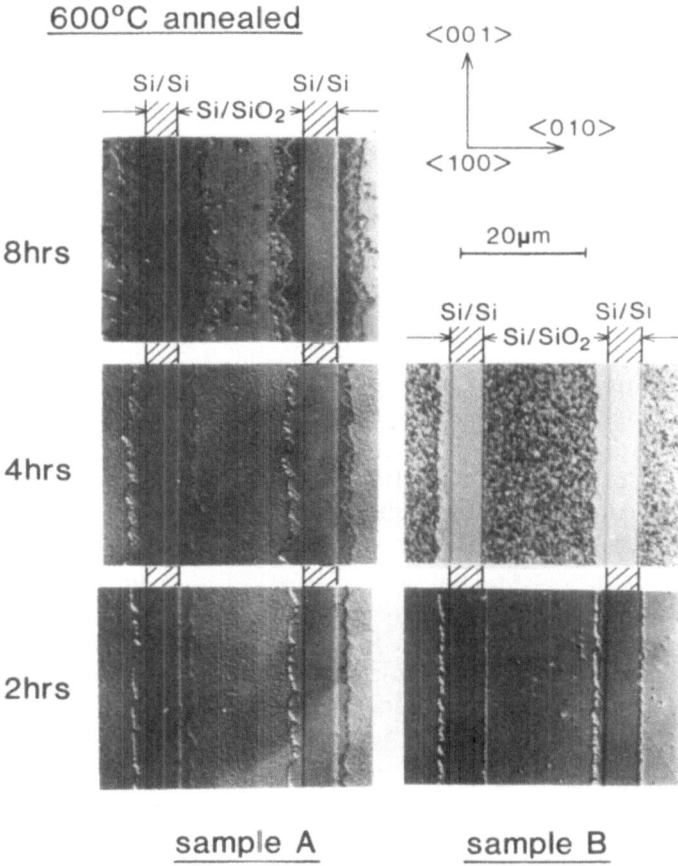

Fig. 10. Nomarski optical micrographs for the samples A and B annealed at 600°C for 2 to 8hours and etched in Wright etchant. Thickness of the SiO₂ film is about 40nm.

A grow by L-SPE from the both sides of the bare Si regions onto the SiO₂ patterns. After 8hours annealing, formation of polycrystalline islands was observed in the remaining a-Si film and the L-SPE was stopped by these polycrystalline grains. The maximum L-SPE length along the ⟨010⟩ direction in this sample was 5–6μm after 600°C-10hour annealing. We can also see from Fig. 10 that L-SPE occurs even in the samples B. This fact indicates that L-SPE occurs even in the room-temperature-evaporated a-Si films when the underlying SiO₂ film is thin enough. However, we can see at the same time that the L-SPE in the samples B is stopped in the 4hour-annealed sample and the maximum L-SPE length is less than 2μm which is much shorter than that of the sample A. That is, the roughly etched surface of the 4hour-annealed sample indicates that the random crystallization time of this sample is less than 4hours. The actual time at 600°C revealed from the detailed investigation was about 2.5hours. Such a short random crystallization time indicates that the stress in the room-temperature-evaporated films is not fully released even if the SiO_2 film is as thin as $\simeq 40$nm. The total L-SPE length in the Si/SiO_2 regions is even reduced because of the fact that the L-SPE was originated only at the left-hand sides of the Si/Si regions. Such one-sided growth is considered due to the evaporation from a slightly oblique direction. In fact, the growth from both sides was observed when the film was deposited by the near-normal-evaporation. However, no difference of the L-SPE length was observed in this case.

We also tried L-SPE of the room-temperature-evaporated a-Si films with the patterns of even thinner ($\simeq 15$nm) SiO_2 films, however, no remarkable improvement was observed. In addition to that, effect of ion beam bombardment to the room-temperature-evaporated a-Si film was examined, i.e., some of the samples B were implanted with Si^+ ions, however, no remarkable improvement was observed. This result indicates that neither the densification of the a-Si films nor the modification of the a-Si/SiO_2 interfaces by ion beam bombardment is so effective to improve the L-SPE charateristics. From these results, we can conclude that use of the dense a-Si films is essentially important for L-SPE, though small L-SPE areas may be obtained even in the porous a-Si films when the thickness of the underlying SiO_2 film is thin enough.

4.4 *Discussions*

The L-SPE growth rates at 600°C measured along the $\langle 001 \rangle$ and $\langle 010 \rangle$ directions in the dense a-Si films on $\simeq 40$nm thick SiO_2 patterns were derived from the micrographs in Fig. 10 and the related ones. The result shown in Fig. 11 has revealed that the L-SPE growth along the $\langle 011 \rangle$ direction starts after less than 30min delay period and continues for about 8hours with a constant growth rate of about 1.2×10^{-8} cm/s. It has also revealed that (1) the L-SPE along the $\langle 010 \rangle$ direction starts almost immediately after the annealing is started, (2) its initial growth rate is higher than that of the $\langle 011 \rangle$ growth, though it gradually decreases to 1.2×10^{-8} cm/s within 3 to 4hours, and (3) the growth continues for 6 to 7 hours with this lowered rate. The rate at the beginning of the $\langle 010 \rangle$ L-SPE can be estimated from the slope of the tangent line at (L-SPE length) = 0 as 3.1×10^{-8} cm/s. These L-SPE rates on the $\langle 011 \rangle$ and $\langle 010 \rangle$ directions are several times lower than the reported V-SPE rates on the (110) and (100) planes, respectively. This fact suggests that the growth fronts in the $\langle 011 \rangle$ and $\langle 010 \rangle$ L-SPE are different from the planes expected in Fig. 2, though some other factors such as defects may also decrease the growth rates.

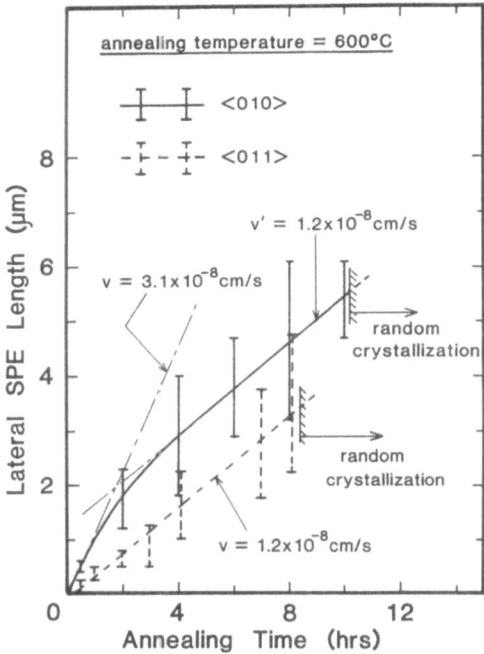

Fig. 11. Plot of the L-SPE length of the samples A along the ⟨010⟩ and ⟨011⟩ directions onto ≃40nm thick SiO₂ patterns as a function of the annealing time at 600°C.

Recently, Kunii *et al.* [16] has shown that the (111) and (110) facets
are formed during the L-SPE growth along the ⟨011⟩ and ⟨010⟩ directions, respectively, as shown in Fig. 12. First, we consider about the ⟨011⟩
growth. In this case, the (111) facet will be formed and its normal has
an angle of 35° to the ⟨011⟩ axis. Though more complicated facets may
be formed in practice, the approximate L-SPE rate along the ⟨011⟩ direction (v(L-SPE⟨011⟩)) is given by

$$v(\text{L-SPE}\langle 011\rangle) = v(\text{V-SPE}(111)) \times \sec 35° \qquad (2)$$

where v(V-SPE(111)) is the V-SPE rate on the (111) plane which has been
reported for self-implanted a-Si films as 0.55 or 1.5×10^{-8} cm/s at 600°C.
In these two V-SPE rates, the latter is thought to be the rate accelerated
by formation of micro-twins.[17] And such micro-twin-accelerated growth
will lead to the defective grown layer and the rough growth front. In
fact, high-density twins and other defects in the L-SPE grown region and
the rough growth front were observed by TEM (Fig. 9). So, we used this
higher V-SPE rate in eq.(2) and obtained a value of 1.8×10^{-8} cm/s as

(a)

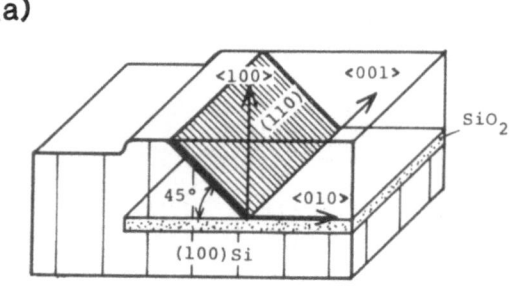

(b)

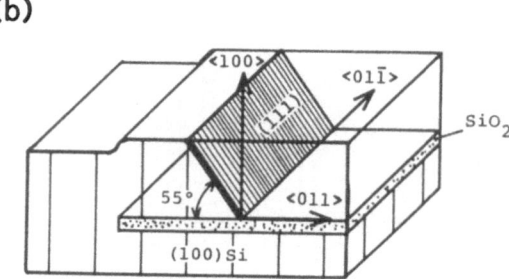

Fig. 12.   (a) The (110) facet formed during the L-SPE along the ⟨010⟩ direction. (b) The (111) facet formed during the L-SPE along the ⟨011⟩ direction. These facets are not normal to the sample surfaces, however, they are parallel to the pattern edges.

the theoretical ⟨011⟩ L-SPE rate. This value is somewhat higher than the measured value ($\simeq 1.2 \times 10^{-8}$ cm/s). However, considering about the facts that the growth rates of evaporated a-Si films may be somewhat lower than those of self-implanted films[9,10] and that the (111) SPE rate may be affected by the densities of the micro-twins, we can say that the L-SPE along the ⟨011⟩ direction is characterized by the growth on the (111) facet plane.

Next, we can see from Fig. 12 that the ⟨010⟩ L-SPE rate with the (110) facet is approximately given by

$$v(\text{L-SPE}\langle 010\rangle) = v(\text{V-SPE}(110)) \times \sec 45° \qquad (3)$$

where $v(\text{V-SPE}(110))$ is the V-SPE rate on the (110) plane which has been reported for self-implanted a-Si films as 3.3×10^{-8} cm/s.[7] Then, we obtained

$$v(\text{L-SPE}\langle 010\rangle) = 4.7 \times 10^{-8} \text{ cm/s}. \qquad (4)$$

This rate should correspond to that at the beginning of the growth

$(\simeq 3.1 \times 10^{-8}$ cm/s). However, the decrease of the rate during the growth in this direction can not be explained by the (110) facet formation.

We presume that the lowered rate $(\simeq 1.2 \times 10^{-8}$ cm/s) is related to the growth on the (111) facet plane. In other words, we presume that the facet at the growth front rotated from the (110) to the (111) plane during the $\langle 010 \rangle$ L-SPE and that the rate was lowered after the rotation because the growth rate on the (111) plane is lower than that on the (110) plane. Though flat (111) facets parallel to the pattern edges can not be formed in this sample configuration, folded (111) and (11$\bar{1}$) facets can be formed as shown in Fig. 13. In fact, we can see zigzag growth fronts in the Nomarski micrographs (Fig. 10) for the sample A annealed at 600°C for 4hours, which can be thought to be the intersects of such (111) facets and the sample surface. The similar zigzag lines can also be seen from the micrograph in the midst of the laterally grown areas for the sample A annealed at 600°C for 8hours. Such zigzag lines can be thought to be the delineation of some kinds of defects lined on the (111) facets. From this micrograph, we can also see that the growth fronts are extremely roughened and the density of the defects are increased after the decrease of the growth rate to $\simeq 1.2 \times 10^{-8}$ cm/s. Further, TEM observation revealed that the $\langle 010 \rangle$ L-SPE grown areas contained high-density twins (Fig. 8). These facts also suggest that the facet rotation to the (111) plane and the micro-twin-accelerated growth on this plane occurred during the growth along the $\langle 010 \rangle$ direction. The (111) facet formation during SPE growth has also been found in the "thin foil annealed (TFA) specimen"[18] and even in the conventional implanted a-Si layers on the (100)Si substrates.[19] As can be seen from Fig. 13, the $\langle 010 \rangle$ growth rate with the (111) facets $(v'(\text{L-SPE}\langle 010 \rangle))$ can be given by

$$v'(\text{L-SPE}\langle 010 \rangle) = v(\text{V-SPE}(111)) \times \sec 45° \times \sec 35°$$
$$= 2.6 \times 10^{-8} \text{ cm/s} \qquad (5)$$

where we used the faster V-SPE rate on the (111) plane. This value is

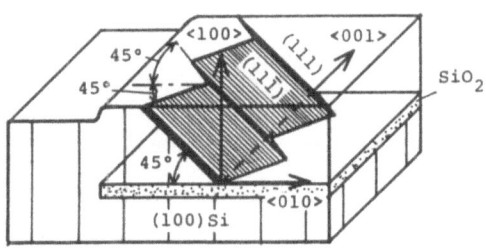

Fig. 13. The (111) and (11$\bar{1}$) facets formed during the L-SPE growth along the $\langle 010 \rangle$ direction. These facets are not parallel to the pattern edge.

somewhat higher than the measured value ($\simeq 1.2 \times 10^{-8}$ cm/s). However, some over-estimation may be included in the theoretical value since the actual amorphous/crystal interfaces are highly irregular and the (111) growth rate depends on the density of the micro-twins.

5. Conclusions

We investigated the lateral solid phase epitaxy of evaporated a-Si films along the $\langle 011 \rangle$ and $\langle 010 \rangle$ directions and found that the dense a-Si films formed by evaporation on heated substrates and subsequent amorphization by Si^+ ion implantation grew laterally onto the SiO_2 patterns. The maximum L-SPE length of 5–6μm onto the $\simeq$40nm thick SiO_2 patterns was obtained along the $\langle 010 \rangle$ direction after 10hour-annealing at 600°C, though the growth front was not smooth and the crystalline quality of the grown areas was not so good. It was found that the growth along this direction was characterized by the growth on the (110) and (111) facet planes and its rate was several times lower than that of V-SPE on the (100) plane.

Acknowledgments
The authors gratefully acknowledge useful discussions with Mr. T.Asano, Mr. N.Natsuaki, and Dr. M.Miyao. This work is partially supported by 1983 Grant-in-Aid for Scientific Research (A) (No.56420024) and 1983 Grant-in-Aid for Special Project Research (No.58109002) from the Ministry of Education, Science and Culture of Japan.

REFERENCES

1) H.M.Lam, Z.P.Zobczak, R.F.Pinizzotto, and A.F.Tasch,Jr.: IEEE Trans. Electron. Devices **ED-29** (1982) 389.
2) K.Shibata, T.Inoue, T.Takigawa, and S.Yoshii: Appl. Phys. Lett. **39** (1981) 645.
3) J.C.C.Fan, B-Y.Tsaur, R.L.Chapman, and M.W.Geis: Appl. Phys. Lett. **41** (1982) 186.
4) Y.Ohmura, Y.Matsushita, and M.Kashiwagi: Jpn. J. Appl. Phys. **21** (1982) L152.
5) Y.Kunii, M.Tabe, and K.Kajiyama: J. Appl. Phys. **54** (1983) 2847.
6) H.Ishiwara, H.Yamamoto, S.Furukawa, M.Tamura, and T.Tokuyama: Appl. Phys. Lett. **43** (1983) 1028.
7) L.Csepregi, E.F.Kennedy, J.W.Mayer, and T.W.Sigmon: J. Appl. Phys. **49** (1978) 3906.
8) A.Lietoila, A.Wakita, T.W.Sigmon, and J.F.Gibbons: J. Appl. Phys. **53**: (1982) 4399.
9) G.L.Olson, S.A.Kokorowski, J.A.Roth, and L.D.Hess: *Proc. Symp. Laser-Solid Interactions and Transient Thermal Processing of Materials, Bonston, 1982*, ed. J.Narayan, W.L.Brown, and R.A.Lemons (North-Holland, New York, 1983).
10) Y.Shiraki: Shingakugiho **CPM80-34** (1980) (in Japanese).
11) N.A.Blum and C.Feldman: J. Non-Cryst. Solids. **11** (1972) 242.
12) K.Zellama, P.Germain, S.Squelard, J.C.Bourgoin, and P.A.Thomas: J. Appl. Phys. **50** (1979) 6995.
13) J.A.Roth, S.A.Kokorowski, G.L.Olson, and L.D.Hess: *Proc. Symp. Laser and Electron-*

Beam Interactions with Solids Boston, 1981, ed. B.R.Appleton and G.K.Celler, (North-Holland, New York, 1982) p169.

14) S.Saitoh, T.Sugii, H.Ishiwara, and S.Furukawa: Jpn. J. Appl. Phys. **20** (1981) L130.

15) M.Tamura, T.Tokuyama, H.Yamamoto, H.Ishiwara, and S.Furukawa: to be published in *Proc. US-Japan Seminar on Solid Phase Epitaxy and Interface Kinetics, Oiso, 1983.*

16) Y.Kunii and K.Kajiyama: to be published in *Proc. US-Japan Seminar on Solid Phase Epitaxy and Interface Kinetics, Oiso, 1983.*

17) R.Drosd and J.Washburn: J. Appl. Phys. **53** (1982) 397.

18) R.Drosd and J.Washburn: J. Appl. Phys. **51** (1980) 4106.

19) J.Narayan: J. Appl. Phys. **53** (1982) 8607.

Silicon-on-Insulator: Its Technology and Applications, edited by S. Furukawa, pp. 209–230.
© KTK Scientific Publishers, Tokyo, 1985.

FORMATION OF A SILICON-ON-INSULATOR STRUCTURE BY SOLID-PHASE EPITAXY

Y. KUNII, M. TABE, and K. KAJIYAMA

Electrical Communication Laboratories, Nippon Telegraph and Telephone Public Corporation, Atsugi, Kanagawa 243-01, Japan

Abstract A silicon-on-insulator (SOI) structure was formed by solid-phase epitaxy (SPE) of chemical vapor deposited (CVD) amorphous silicon (a-Si). Essential conditions for SPE were fulfilled utilizing a "clean-CVD" process. In-reactor cleaning removed all interface layers between the a-Si film and Si substrate. Low deposition temperature and high deposition rate reduced micro-crystallites and foreign atoms in the a-Si film. The a-Si/SPE-layer facet was investigated and explained with an atomistic model. A principle for further improvement in lateral SPE is discussed.

1. Introduction

Solid-phase epitaxy (SPE) of chemical vapor deposited (CVD) amorphous silicon (a-Si) is important in applications to LSIs. For example, lateral SPE produces a silicon-on-insulator (SOI) structure with good surface morphology, even if no cap is applied.[1,2] On the contrary, in lateral liquid-phase epitaxy, good surface morphology is difficult to obtain without lengthy encapsulation processing. In addition, the present process supplies low cost SOI substrates. The CVD process gives higher throughput than in-vacuum deposition. Furnace annealing promises good reproducibility and batch processing of many wafers. Also, using a thick CVD amorphous silicon film, SPE growth can be clearly investigated by conventional scanning electron microscope (SEM) observation of sample cross-sections after chemical etching.

A new CVD process, called clean-CVD,[3] provides a suitable a-Si film for SPE growth. It excludes all interface layers between the deposited film and the substrate, in addition to micro-crystallites or any foreign atoms in the deposited film. These hurdles had been the main drawbacks hindering the development of SPE of CVD a-Si.

This paper presents the following items:

1) The clean-CVD surface cleaning process. The cleaning process consists of two etching steps: native oxide etching with H_2; and Si light etching with HCl to prevent adsorption of foreign atoms that may obstruct SPE.

2) The clean-CVD deposition temperature and deposition rate conditions. A low deposition temperature prevents micro-crystalline formation in the deposited a-Si film. Micro-crystallites lead to random crystallization inside the film during heat treatment. A high deposition rate hinders foreign atom inclusion. Foreign atoms slow down the SPE growth rate.

3) Lateral SPE and facet formation. A facet is formed at the interface between the a-Si film and SPE layer during SPE growth near SiO_2. The facet formation mechanism is discussed with an atomistic model.

4) Further improvements in lateral SPE. A guiding principle to improve SOI width and crystallinity is discussed.

2. *Experimental Procedure*

Substrates were p-type Si(100) wafers with a 3-inch diameter, having resistivities in the 1–10 ohm·cm range. To compare with lateral SPE, bare wafers were used for the vertical SPE experiment. For the lateral SPE experiment, an SiO_2 stripe was formed on the substrate. The wafers were thermally oxidized at 1100°C in a dry O_2 atmosphere to form a 0.1–0.2 μm thick SiO_2 film. Next, stripe patterns of SiO_2 (2–500 μm wide stripes and 2–20 μm openings) were formed parallel to the [001] and [011] directions, using a conventional photolithography process.

The a-Si film was deposited on the substrate by the following clean-CVD process,[3] which is explained below in Section 3. 1 and 2. Figure 1 shows the process for lateral SPE. After conventional chemical cleaning of the substrate surface, samples were loaded into an rf-inductive CVD reactor. First, the native oxide layer at the open Si surface was removed by H_2 gas etching at 1100°C. Second, while lowering the temperature to the low deposition temperature, the cleaned surface was protected from adsorbing impurities (oxygen, etc.) by performing HCl gas etching.

After in-reactor cleaning, the 0.5–2.0 μm thick a-Si film was deposited on the substrate with a source gas, SiH_4 (partial pressure $= 6 \times 10^{-2}$ atm), and a carrier gas, N_2 at 580°C. For the lateral SPE experiment, the deposition rate of 0.2 μm/min was used.

Deposited films were furnace-annealed in a dry N_2 atmosphere at 525–1150°C. During annealing, SPE growth first took place in the vertical direction and then in the lateral direction over the SiO_2 stripe. After furnace-annealing, the samples were cleaved, and etched by th Wright solution.[4] The growing interface between a-Si and crystalline Si (c-Si) was then observed by SEM.

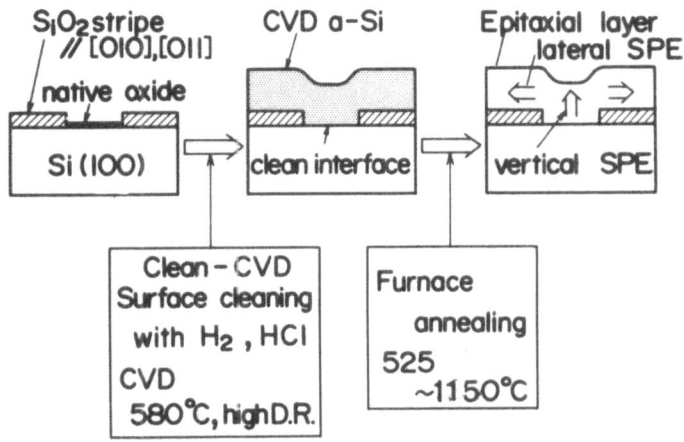

Fig. 1.   Process diagram for lateral SPE.

3.   Results

3.1   Surface cleaning process in clean-CVD

Substrate surface cleaning is essential to SPE of deposited a-Si films. Interface layers block epitaxial growth if cleaning is insufficient. In a conventional Si vapor-phase epitaxy, pure H_2 and/or dilute HCl is used for vapor etching in order to prepare clean surfaces. While etching rates of Si with dilute HCl gas had been already investigated in the temperature range of 1100–1300°C,[5,6] the a-Si deposition temperature is below 650°C.[7] In this low temperature range, etching characteristics of HCl and H_2 had not been reported. Accordingly, the authors decided to investigate the Si and SiO_2 etching rates with H_2 and HCl in the temperature range between 550°C and 1100°C.[3]

The temperature dependence of SiO_2 etching rates with pure H_2 and dilute HCl (0.5% with H_2) is shown in Fig. 2. Both etching rates coincidently varied from 4 Å/min at 1100°C to an undetectably small value below 900°C. Consequently, for etching SiO_2, H_2 was effective above 1100°C, but HCl was not effective in the temperature range of 550–1100°C.

The temperature dependence of Si etching rates with pure H_2 and dilute HCl (0.1% or 0.5% with Ar) is shown in Fig. 3. In the etching rate measurement, the authors used low pressure CVD polycrystalline Si films on thermally oxidized Si substrates. With a $Si/SiO_2/Si$ structure, surface Si layer thickness was easily measured using conventional optical interference equipment (Nanospec). In order to obtain the actual etching rate of Si, the native oxide must be removed before Si etching.

Y. Kunii *et al.*

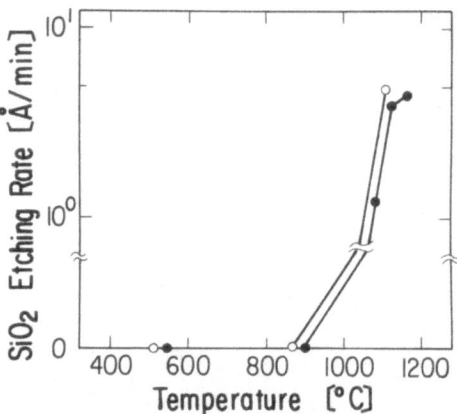

Fig. 2. Temperature dependence of SiO_2 etching rates of with H_2 (solid circle) and dilute HCl (open circle).

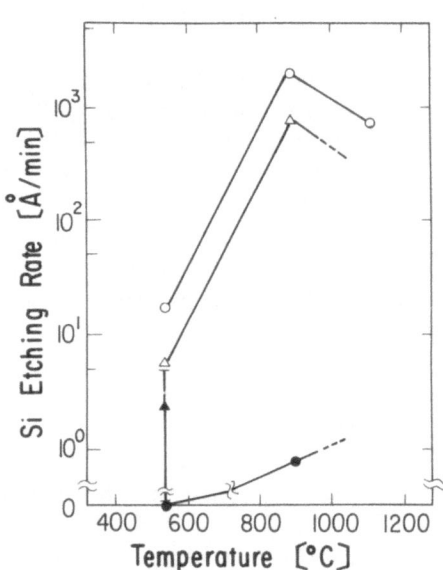

Fig. 3. Temperature dependence of Si etching rates with H_2 (solid circle) and with dilute HCl (open circle: 0.5 % with Ar, and open triangle: 0.1 % with Ar), after H_2 vapor etching of native oxide. Without prior H_2 vapor etching, Si etching rate with dilute HCl (0.1 % with Ar) lowered to solid triangle from open triangle.

    The native oxide on polycrystalline Si was removed by H_2 vapor etching at 1100°C for a few minutes. The etching rate was about 4 Å/min, as shown in Fig. 2. The Si etching rate with dilute HCl (0.1% with Ar) was about 5 Å/min, even at the low temperature of 550°C. Without prior H_2 etching, the Si etching rate was about half that with prior H_2 etching. This is because the native oxide restrains HCl vapor etching.

    Under optimized cleaning conditions, CVD a-Si crystallizes in SPE. The cleaning conditions were selected on the basis of whether SPE took place or not, because direct surface cleanliness measurements such LEED or AES in a ultra-high vacuum were difficult to carry out in the CVD reactor used. After the previously-mentioned chemical etching and some vapor eching, a thin a-Si film was deposited at 550°C with a SiH_4 partial pressure of 6×10^{-3} atm. The film thickness was 500 Å or less, to reduce the crystallizing period. The films were then heated at 600°C for periods of between 15 minutes and 24 hours. Surface crystallinity was investigated by reflection electron diffraction (RED) to examine whether SPE took place or not.

    Utilizing the optimized cleaning process shown in Fig. 4, a single crystalline surface was obtained after 15 minutes of heating. However, SPE did not take place when the H_2 vapor etching temperature was below 900°C. Also, it did not occur when the HCl vapor etching was stopped halfway between H_2 vapor etching and the deposition. Therefore, both SiO_2 etching at high temperature and Si etching up to just before Si deposi-

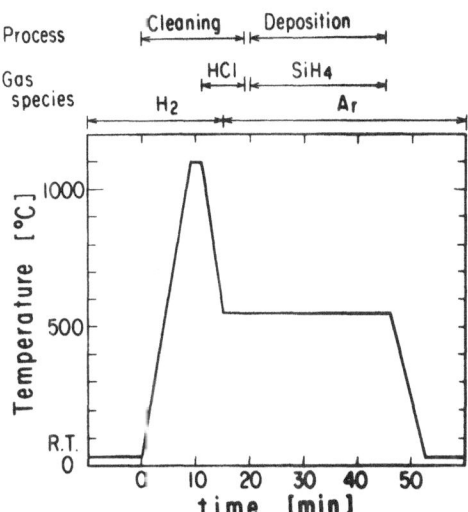

Fig. 4. CVD process diagram for SPE. Cleaning process consists of SiO_2 etching with H_2 at 1100°C and subsequent Si etching with dilute HCl (0.1 % with H_2 or Ar) up to deposition. Deposition process is SiH_4 pyrolysis in Ar carrier gas at 550°C.

tion were necessary to avoid any interface layer formation between the substrate and the deposited film.

3.2 Clean-CVD amorphous Si deposition conditions

Introduced first is the deposition rate and characteristics of the CVD a-Si,[3] as there has been only a few reports about a CVD a-Si.[7-10]

Si films were deposited by pyrolysis of SiH_4 with Ar carrier gas at 500–700°C. The carrier gas was Ar and not H_2 as the Si deposition rate with H_2 carrier gas is one order of magnitude lower than that with Ar carrier gas at the present experimental temperatures.[11] Temperature dependence of the Si deposition rate at SiH_4 partial pressure of 6.0×10^{-4}–1.2×10^{-2} atm in Ar is shown in Fig. 5. Activation energy E_0 was 2.2 eV in the low temperature range. Hereafter, a SiH_4 partial pressure of 6×10^{-3} atm lower temperatures ($<600°C$) and one of 6×10^{-4} atm at higher temperatures ($>600°C$) were mostly used, in order to get 1000–3000 Å thick films within a few tens of minutes of deposition time.

The characteristics of deposited Si are also essential to SPE. If any crystalline part exists in the deposited Si, random crystallization will take place during the heat treatment. The deposition condition dependence of CVD a-Si characteristics were examined by transmission electron microscope (TEM) and transmission electron diffraction (TED).

TEM and TED reveal crystal size and orientation. TEM and TED photographs of Si films deposited at various temperatures are shown in Fig. 6. At 700°C, deposited films were polycrystalline, with TED pattern

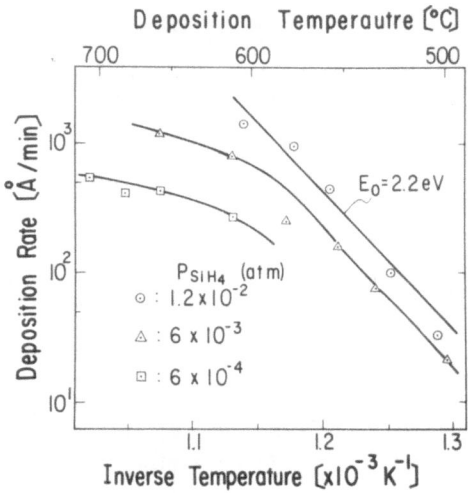

Fig. 5. Arrhenius plot of temperature dependence of deposition rate in SiH_4/Ar. Activation energy $E_0 = 2.2$ eV in lower temperature range.

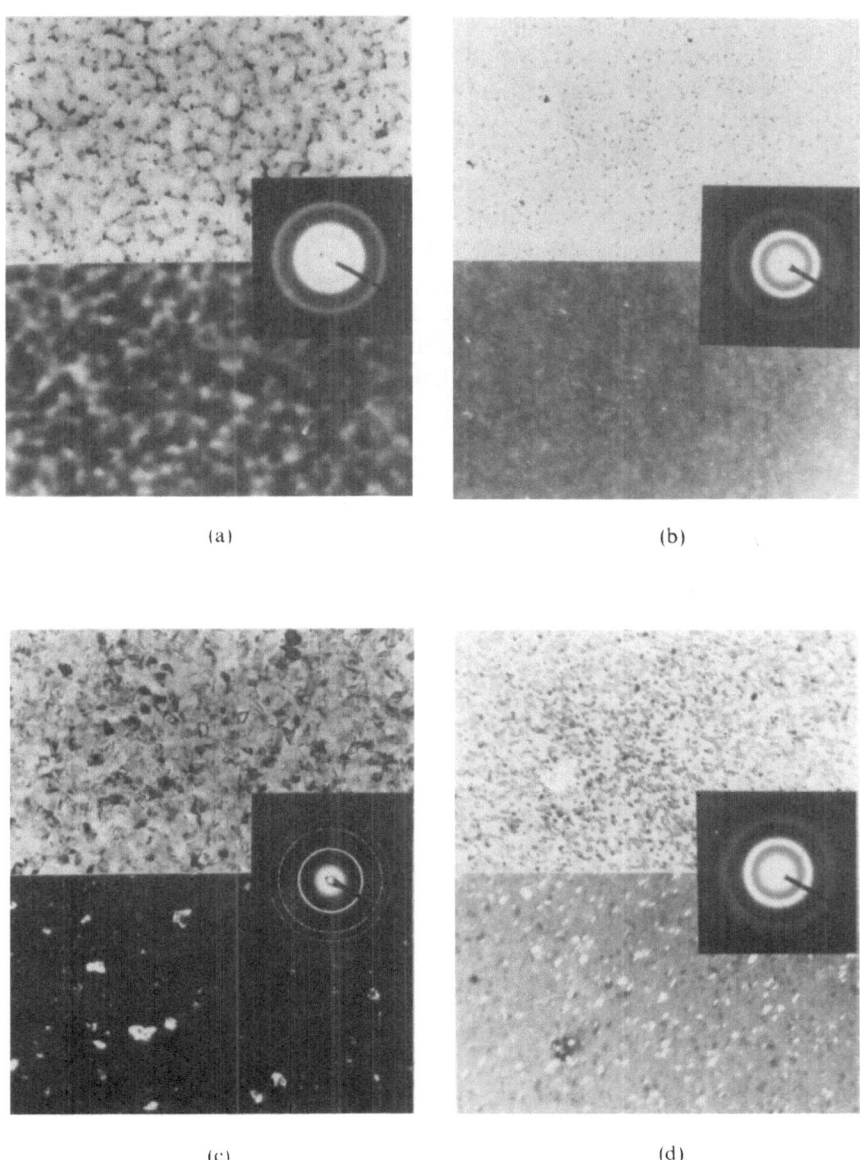

(a)   (b)

(c)   (d)

Fig. 6.   TEM and TED photographs of CVD Si films. Bright-field (upper) and dark-field (lower) TEM photographs, and corresponding selected-area TED patterns (inset). Deposition temperatures: (a) 500°C, (b) 550°C, (c) 650°C, and (d) 700°C.

being ringed and TEM showing polycrystalline grains. Below 650°C, deposited films were "amorphous", with halo TED patterns. However, in the temperature range around 650°C, many micro-crystallites were observed in both the bright- and dark-field TEM photographs. The micro-crystallites were found within the amorphous phase. Their size was about 100 Å.

Micro-crystallites grow in random orientation during the heat treatment. Therefore, a low deposition temperature is preferable to avoid random grain growth, as described below.

After chemical and optimized vapor eching (Section 3.1), 1000–3000 Å thick a-Si films were deposited at 550–620°C with a SiH_4 partial pressure of 1.2×10^{-4}–3.0×10^{-2} atm. The films were heated at 600°C in dry N_2. The epitaxial layer thickness and crystallinity were examined by channeling Rutherford backscattering (RBS) and reflection electron diffraction (RED).

When deposition temperature was 550°C, the CVD film crystallized epitaxially on the substrate. As is shown in Fig. 7, the epitaxial layer front moved in parallel with the substrate surface within the measurement depth resolution of 150 Å. After 60 minutes of heat treatment, the channeling spectrum of the epitaxial layer was not distinguishable from that

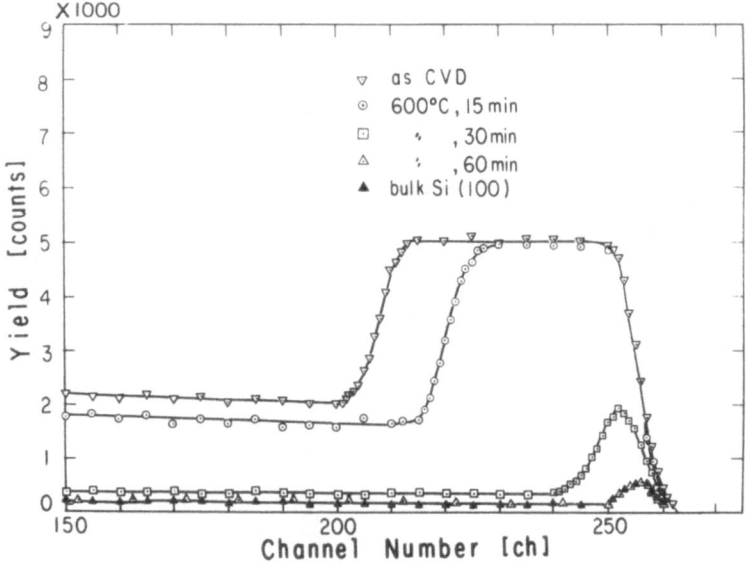

Fig. 7. Rutherford backscattering measurement. [100] channeling spectra change with SPE growth. a-Si film deposited at 550°C by CVD. Probe: He⁺ of 1.54 MeV. SSB detector inclined at 110° with He⁺ probe bean direction to improve depth resolution to 150 Å.

of bulk Si(100); crystalline defects were not observed in either the SPE layer or the interface between SPE layer and the substrate. The RED pattern of the epitaxial layer was also similar to that of bulk Si(100), as shown in Fig. 8.

However, when deposition temperature was much higher than 550°C, the CVD film did not epitaxially crystallize to the substrate. To illustrate, Figure 9 presents crystallization characteristics of 620°C deposited film. The epitaxial front inclined after 60 minutes of heat treatment at 600°C. After 24 hours of heat treatment, the RED pattern consisted of normal reciprocal lattice spots and the extra spots of {111} twins, as shown in Fig. 10.

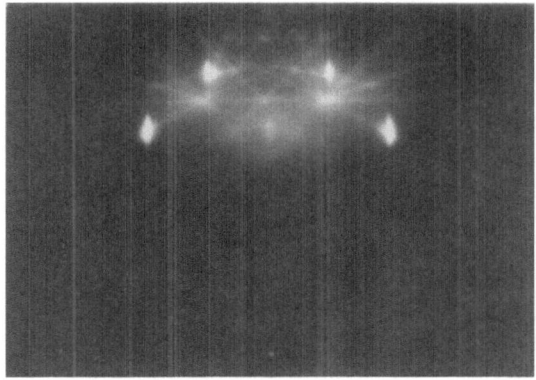

(a)

(b)

Fig. 8.   Reflection electron diffraction pattern. Sample on Si(100) substrate and 60 keV electron beam incident in the [010] direction. (a) Epitaxial layer fabricated by SPE of CVD a-Si. (b) Reference single crystal Si.

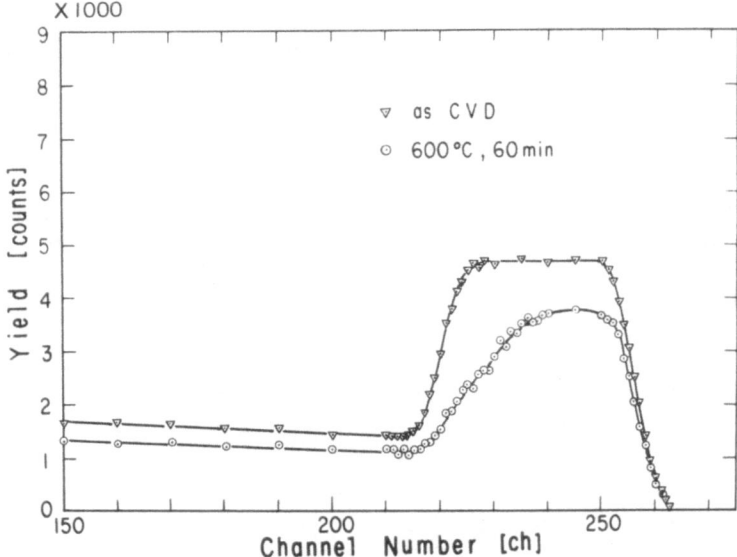

Fig. 9. Change in RBS channeling spectra with partial growth of SPE and random crystallization. a-Si film deposited at 620°C. Measurement conditions same as Fig. 7.

Fig. 10. Reflection electron diffraction pattern (same sample in Fig. 9), after 24 hours heat treatment at 600°C. 60 keV electron beam incident in the [011] direction.

SPE depended on deposition rate as well as deposition temperature. The deposition rate was changed by SiH$_4$ partial pressure. The deposition rate dependence of the SPE growth rate at 600°C is shown in Fig. 11. With a high deposition rate (0.2 μm/min), the epitaxial growth rate was about 370 Å/min. The growth rate was between that of the self-implanted a-Si[12] and that of the evaporated a-Si.[13] The growth rate of CVD a-Si decreased with the deposition rate decrease. The decrease was probably due to the increase of foreign atom inclusion during deposition. Therefore, it can be inferred that a higher deposition rate is desirable for excluding foreign atoms in CVD film.

3.3   Vertical solid-phase epitaxy

Vertical SPE growth rate on a bare Si substrate was carefully investigatd for reference with lateral SPE. After a delay, vertical SPE starts and the epitaxial layer grows at a constant growth rate within the 500 Å measurement depth resolution (Fig. 12 and 13). The SPE layer front moves parallel to the original interface and no facet is formed during vertical SPE. The vertical SPE delay (t_v) decreases with an annealing temperature (T) increase (Fig. 14), and is given by eq.(1)

$$t_v = 5 \times 10^{-11} \min \times \exp(2.0\,\mathrm{eV}/kT). \tag{1}$$

It is speculated that the delay is due to some residual contamination in

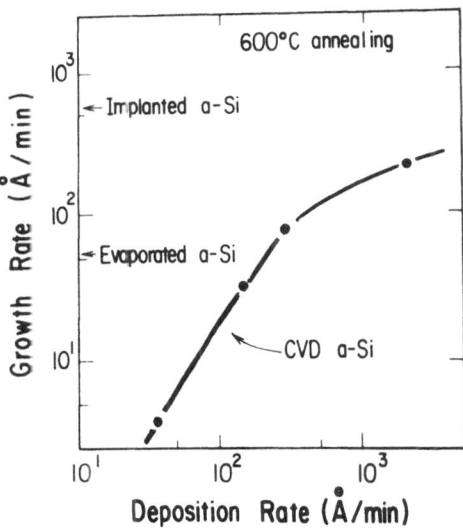

Fig. 11.   CVD deposition rate dependence of SPE growth rate at 600°C. CVD a-Si deposited at 550–580°C. Growth rates of self-implantd a-Si[12] and evaporated a-Si[13] at the same temperature are also shown.

Y. Kunii *et al.*

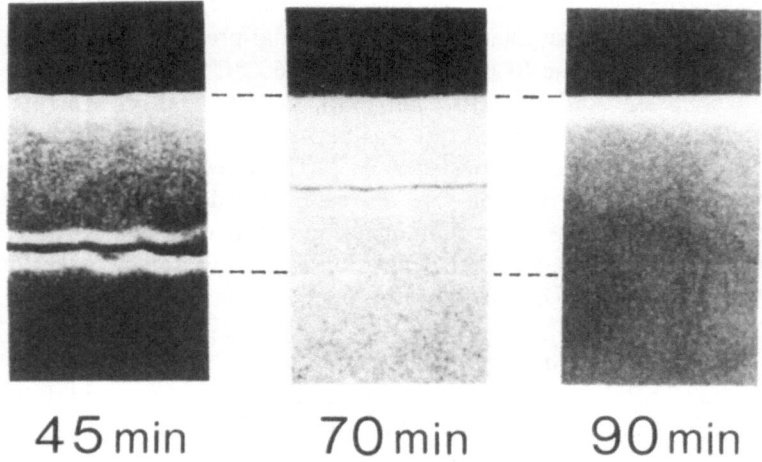

45 min   70 min   90 min

Fig. 12. SEM photographs of sample section during vertical SPE with 575°C annealing for 45 minutes, 70 minutes and 90 minutes.

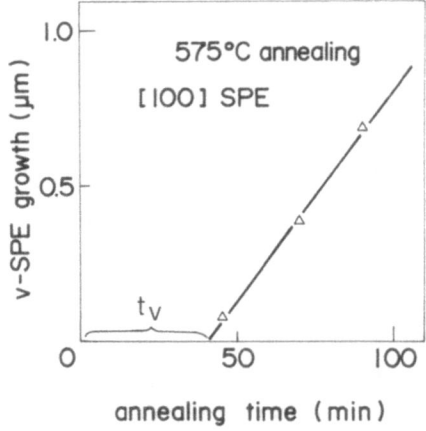

Fig. 13. Vertical SPE growth dependence on annealing time.

the a-Si/substrate-Si interface.

The vertical SPE growth rate increases with an annealing temperature increase (Fig. 15). The growth rate activation energy is 3.2 eV between 525°C and 600°C. The vertical SPE growth rate is given by experimental eq.(2)

$$V_v = 1.2 \times 10^{21} \, \text{Å/min} \times \exp(-3.2 \, \text{eV}/kT). \qquad (2)$$

The activation energy for the present CVD a-Si is larger than that for

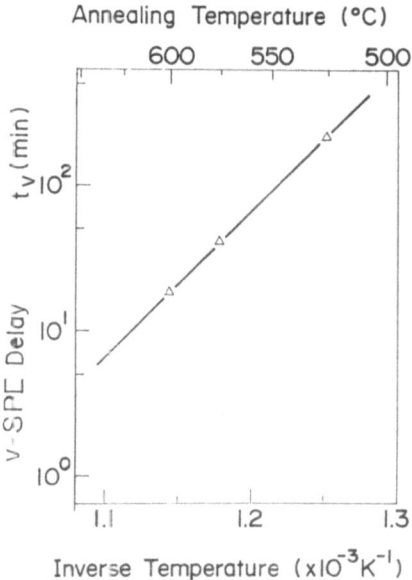

Fig. 14. Arrhenius plot of vertical SPE delay time dependence on annealing temperature.

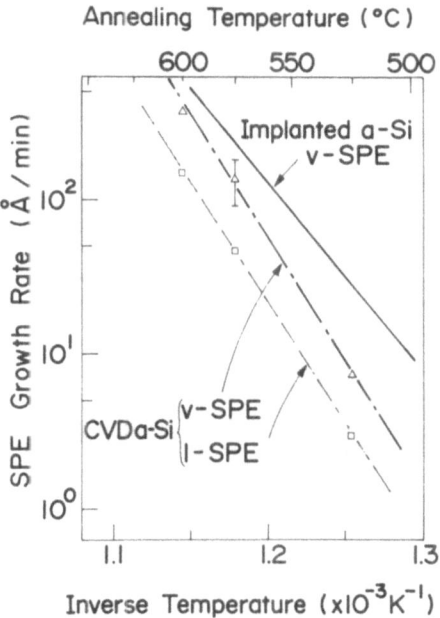

Fig. 15. Arrhenius plot of vertical SPE (v-SPE) and lateral SPE (1-SPE) growth rate dependence on annealing temperature.

implantation-formed a-Si (2.35 eV[12]–2.8 eV[14]), probably because of impurities in CVD a-Si film. With an increase in annealing temperature, CVD a-Si vertical SPE growth rate becomes close to that for implantation-formed a-Si (25% at 525°C and 65% at 600°C). Extrapolating the growth rate to higher temperatures, the CVD a-Si vertical SPE growth rate will reach that of the implantation-formed a-Si somewhere around 630°C.

3.4 Lateral solid-phase epitaxy

When there is a SiO_2 pattern at the substrate surface, the lateral SPE delay (t_L) reaches 2.2 times that of the vertical SPE delay (t_v) for a bare Si substrate at 600°C. The extent of SPE growth is shortened in the area close to the SiO_2 stripe (Fig. 16(a)). This is due to either the increase

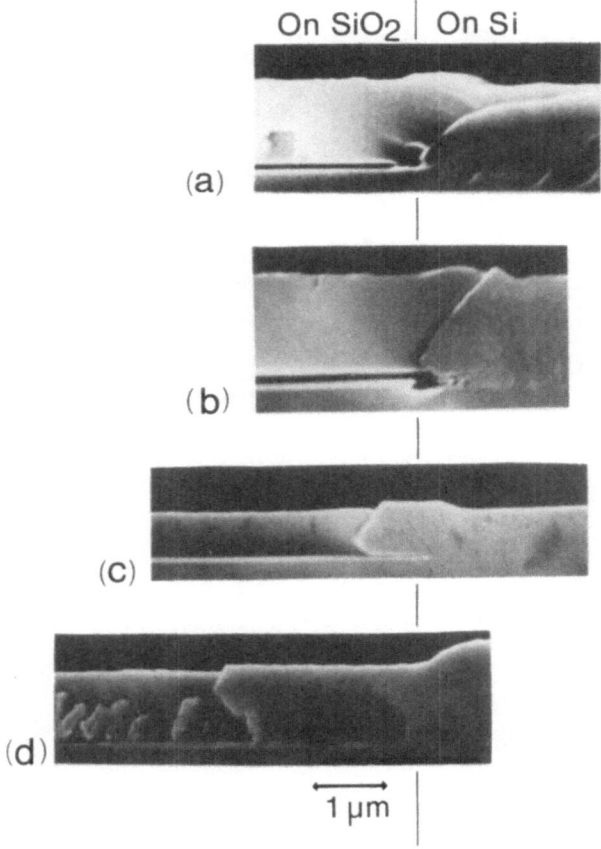

Fig. 16. SEM photographs of sample section during lateral SPE for [001] stripe. (a) 550°C annealing for 22 hr. 600°C annealing for (b) 60–120 minutes, (c) 90 minutes and (d) 180 minutes.

in delay or the decrease in SPE growth rate near the deposited-layer/substrate interface. Either is a result of oxygen contamination from the SiO_2 stripe.

Lateral SPE is strongly affected by the SiO_2 stripe direction to the substrate crystalline orientation. During lateral SPE, a facet is formed at the SiO_2 stripe edge (Fig. 16(b)). The angle between this facet and the (100) plane is 42°–50° for [001] stripe, and is almost equal to the 45° angle between the (100) and (110) planes. Since (110) facet formation is predicted by the atomistic model (Section 4.2). the authors concluded that the facet orientation is (110) for the [001] stripe.

During SPE growth on SiO_2 stripe, the a-Si/c-Si interface forms two facets (Fig. 16(c),(d)). The ($\bar{1}$10) facet is formed at the a-Si/SiO_2 boundary and the (110) facet is formed at the a-Si/surface-native-oxide boundary on [001] stripe. Figure 17 shows a top view of the sample during lateral growth on SiO_2. The a-Si/c-Si interface shows small undulation, but no facet is observed from the top view. This implies that there is no large obstacles in a-Si; therefore SPE proceeds uniformly along SiO_2 stripe.

It should be noted that good surface morphology for SPE-Si on SiO_2 is maintained, although there is no cap layer, except for the native oxide layer, on SPE-Si. This is because lateral epitaxy occurs in the solid-phase and not in the liquid-phase.

The lateral SPE growth rate (V_L) was determined to be 370 Å from

On SiO_2 $\longrightarrow$ $\longleftarrow$ On Si

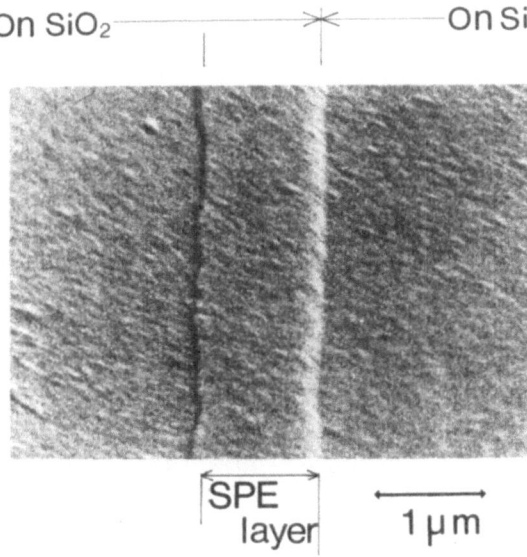

SPE layer   1 μm

Fig. 17. SEM photograph of sample top view during lateral SPE for [001] stripe (600°C annealing for 135 minutes).

the lateral growth length (L) at three different annealing times (t_a) at 600°C. The V_L dependence on annealing temperature was estimated and was given by eq. (3)

$$V_L = 4.7 \times 10^{20}\,\text{Å/min} \times \exp(-3.2\,\text{eV}/kT). \tag{3}$$

The V_L was 40% of the V_v.

Also, an a-Si/c-Si facet is formed during lateral SPE for the [011] stripe (Fig. 18). The angle between the facet and the (100) plane is 49–55°, and is almost equal to the 54.7° angle between the (100) and (111) planes. Comparing this result with the atomistic model, the facet orientation is concluded to be (111). The lateral SPE growth rate on the [011] SiO$_2$ stripe is small (60 Å/min at 600°C). {111} twins are formed densely on SiO$_2$ stripe (Fig. 19). This result also supports the (111) facet formation, because microtwins are easily created on the (111) plane, as pointed out by Drosd and Washburn.[15] The facet formation mechanism is explained by an atomistic SPE model below in Section 4.2.

Figure 20 shows the dependence of single crystalline SOI width on annealing temperature. The SOI width is one of the most important factors for application of this sort. Single crystalline SOI width is the product of the SPE growth rate and growth time. SPE growth begins after a delay, and ends when random crystallization obstructs SPE. The SOI width decreases with an increase of annealing temperature up to 900°C, because the annealing temperature increase enhances random crystallization more than it does SPE growth. The SOI width decreases abruptly to an undetectable value above 900°C, because random crystallization occurs during SPE delay. In the present study, a 5 μm wide SOI area is formed from the SiO$_2$ stripe edge by annealing at 525°C.

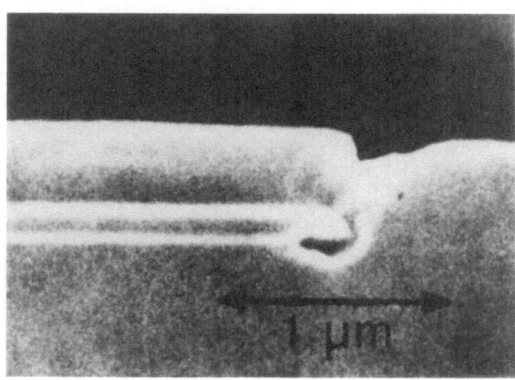

Fig. 18. SEM photograph of sample section during lateral SPE for [011] stripe (600°C annealing for 2 hr).

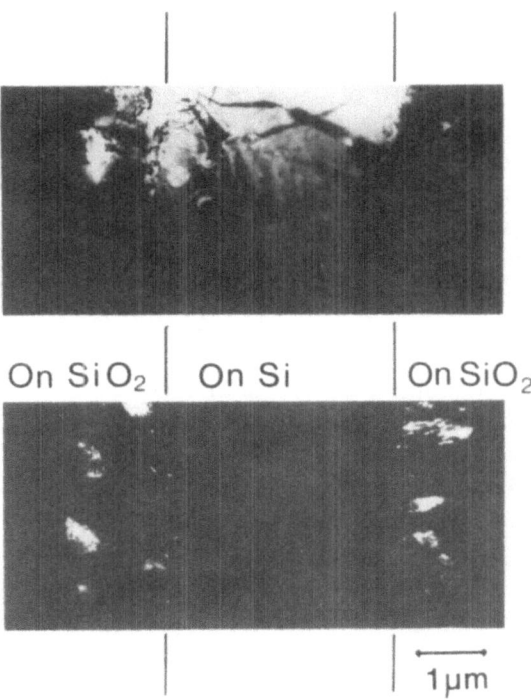

On SiO₂ | On Si | On SiO₂

Fig. 19.   Bright-field (upper) and dark-field (lower) TEM photographs of sample during lateral SPE for [011] stripe (600°C anneal for 2 days).

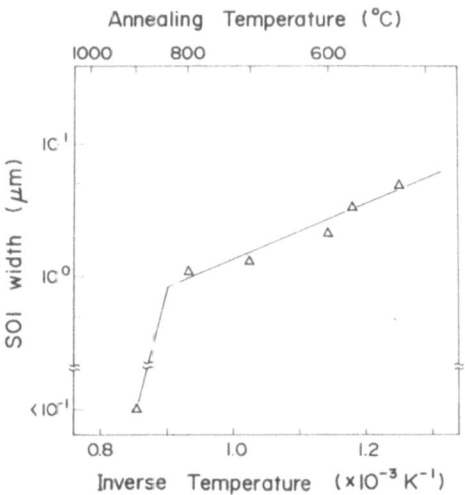

Fig. 20.   SOI width dependence on annealing temperature.

4. Discussion

4.1 Clean-CVD process

In order to attain SPE, two essential conditions are acomplished in the clean-CVD process: removing the interface layer and avoiding micro-crystallite or foreign atom inclusion in the a-Si layer.

First, the native oxide and adsorbed layer should be removed before deposition so that amorphous (a) phase atoms adjoin the crystalline (c) phase. The native oxide layer of 10 Å thickness[16] was removed by H_2 vapor etching at 1100°C within a few minutes in accordance with the SiO_2 etching rate shown in Fig. 2. The adsorbed layer was removed by HCl vapor etching in accordance with the Si etching rate shown in Fig. 3.

Second, it is desirable that a-Si layer includes no micro-crystallites or foreign atoms. Micro-crystallites are effectively eliminated in the a-phase with deposition at low temperatures of 550–580°C. However, the a-Si film deposited at a higher temperature grows epitaxially only during its early growth stage and becomes a crystalline layer with high defect density (Fig. 9 and Fig. 10). This is because the deposited film includes micro-crystallites. As the micro-crystallite size is about 100 Å (Fig. 6), the component atom number is about 10^5. On the other hand, the atom number for critical nucleation at 600°C is about 10^2,[17] and is far smaller than the component atom number of the micro-crystallite. Thus, micro-crystallites grow during heat treatment.

Foreign atoms are reduced in a-Si film with high deposition rate. As mentioned earlier, the SPE growth rate increased with an increase in deposition rate (Fig. 11). This fact suggests that a high deposition rate reduces the incorporation of impurities (oxygen, etc.) which suppress the SPE growth rate.[18] Therefore, both a low deposition temperature and a high deposition rate are required for SPE.

4.2 Atomistic model for facet formation

In this section, a-Si/c-Si facet formation is explained with an atomistic model. A (110) facet is formed during [100] SPE growth bounded by the (010) plane ((010)b-[100]SPE, Fig. 16(b)). A (111) facet is formed during (011)b-[100]SPE (Fig. 18). The facet formation is explained with the Drosd and Washburn's model (D-W model),[15] which assumes that an a-Si atom must form at least two undistorted bonds for SPE growth, and the boundary condition wherein an a-Si atom cannot form undistorted bonds to the boundary.

First, consider the (010)b-[100]SPE case (Fig. 21(a)). If an a-Si atom is at the a-Si/c-Si interface and is distant from the boundary (SiO_2), it forms undistorted bonds with seed c-Si atoms, breaking and distorting the bonds to the other a-Si atoms. However, if an a-Si atom is at the

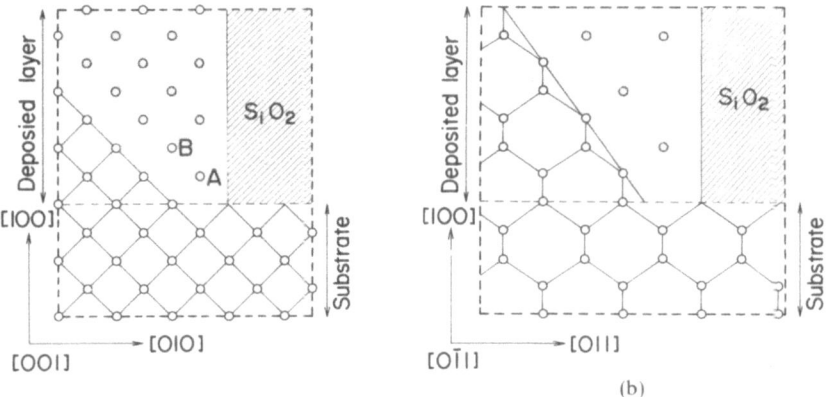

Fig. 21. bounded SPE process based on Drosd-Washburn model[8] for (a) (010)b-[100]SPE
and (b) (011)b-[100]SPE.

a-Si/c-Si interface and is attached to the boundary (atom "A" in Fig.
21(a)), it cannot form two undistorted bonds with seed c-Si atoms. Applying the boundary condition wherein an a-Si atom cannot form undistorted
bonds to the boundary, the a-Si atom remains in a-phase. Successively,
the 2nd layer grows in SPE on the first SPE layer, but recedes from the
boundary by $\sqrt{2}$ of spacing between (110) planes, because atom "B" in
the figure remains in the a-phase. Consequently, a-Si/c-Si interface forms
a (110) facet, as (010)b-[100]SPE goes on. After the facet is formed, face-
surface growth proceeds in SPE, according to its crystalline orientation,
i.e. [110] in this case.

Second, consider the (011)b-[100]SPE case. Similar growth process
occurs and the (111) facet is formed (Fig. 21(b)). During [111] SPE, microtwin formation occurs (Fig. 19), because a twin nucleus can be created
on the {111} seed plane, as easily predicted by the D-W model.[15]

4.3 Normalized lateral solid-phase epitaxy growth rate

In this section, it is shown that normalized growth rates (setting the
growth rate of [100] SPE at 1.00) for the [001] stripe and [011] stripe
are consistent with normalized growth rate dependence on substrate
orientation[12] and secant factor of angle between the growth direction and
facet plane.

As shown in Fig. 22(a), in SPE growth on the [001] SiO$_2$ stripe, a
(110) facet grows in the [110] direction and a ($\bar{1}$10) facet grows in the
[$\bar{1}$10] direction. This results in two {110} facets growing in the [010] direction. The atomistic description of the growth is as follows (Fig. 22(b)):
(1) A two atom cluster nucleation on (110) or ($\bar{1}$10) facet. (2) Atom

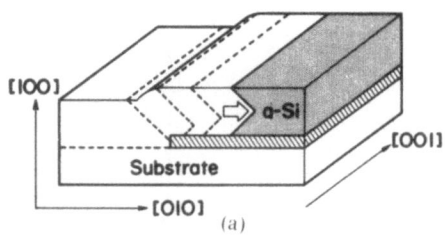

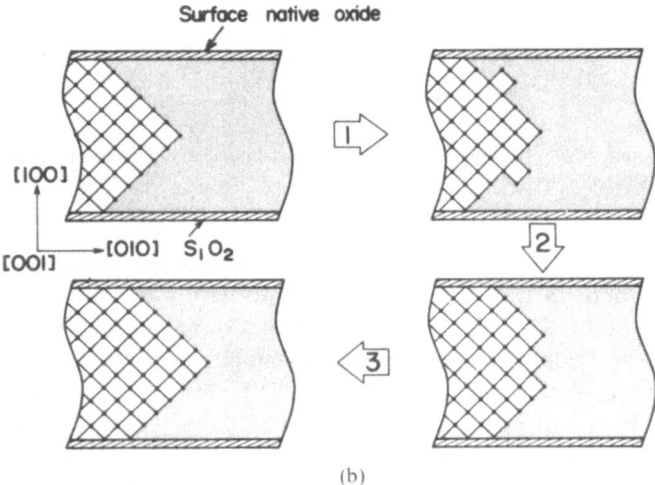

Fig. 22. (100)b-[010]SPE. (a) Diagram. (b) Atomistic description.

cluster growth in [$\bar{1}$10] or [110] direction resulting in the small (010) facet. (3) Growth in [010] direction which annihilates the (010) facet. Therefore, the growth rate for [001] stripe is limited by [110] SPE and is given by eq. (4),

$$V(\text{on [001] stripe}) = V([110]\,\text{SPE}) \times \sec 45°. \tag{4}$$

Since [110] SPE growth rate has not been investigated for a clean-CVD a-Si, the authors assume that the normalized growth rate dependence on substrate orientation for implantation-formed a-Si ($V([110]\,\text{SPE}) = 0.33$)[12] can be applied in this case. Thus, the normalized growth rate for [001] stripe is given by eq. (5),

$$\begin{aligned}
V(\text{on [001] stripe}) &= V([110]\,\text{SPE}) \times \sec 45° \\
&= 0.33 \times \sec 45° \\
&= 0.47.
\end{aligned} \tag{5}$$

This calculated value agrees well with the experimental value of 40% (Section 3.4).

SPE growth on [011] SiO_2 stripe is described in Fig. 23. Applying data for implantation-formed a-Si ($V([111] SPE) = 0.15$ after 1000 Å SPE growth),[12] the normalizd growth rate for the [011] stripe is given by eq.(6),

$$V(\text{on}[011]\text{stripe}) = V([111]SPE) \times \sec 35.3°$$
$$= 0.15 \times \sec 35.3°$$
$$= 0.18. \qquad (6)$$

This calculated value agrees well with the experimental normalized growth rate value $((60 \text{ Å/min})/(370 \text{ Å/min}) = 0.16$ at 600°C, Section 3.4).

4.4 Further improvement in lateral SPE

Further improvement in SOI width and crystallinity will be realized through refining clean-CVD and annealing conditions. Refining the surface cleaning process will decrease the SPE delay and will improve SPE layer crystallinity. In addition, as mentioned above, reduction of microcrystallites and foreign atoms in a-Si layer increases the vertical and lateral SPE growth rate.

Furthermore, irradiation of high-energy heavy-ion (As) beam increases vertical growth rate nominally 10^6 times at low temperature of 290°C. This annealing method is called HIBA.[19] The growth rate increase is not due to the Ar doping effect, but is the result of high-energy heavy-ion beam. HIBA's crystal growth mechanism is different from SPE's mechanism. If HIBA is applied to lateral SPE, SOI width will widen because random crystallization is negligible at this low temperature.

5. Conclusion

Results and discussion are summarized as follows:

1) Using the clean-CVD technique, the following SPE conditions were accomplished. First, a clean interface between the substrate and amor-

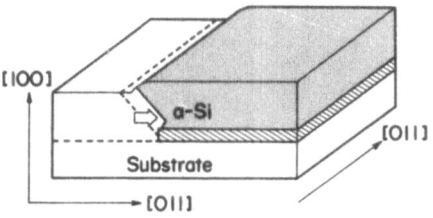

Fig. 23. (100)b-[011]SPE diagram.

phous silicon film was achieved. Second, a reduction in micro-crystallites and obstructive impurities in amorphous silicon film was obtained.

2) Facet formation occurs during lateral SPE. For example, a (110) facet is formed for [001] SiO_2 stripe and a (111) facet is formed for [011] SiO_2 stripe on a (100) silicon substrate. Facet formation during lateral SPE is explained by the two bonds per atom requirement and the boundary condition.

3) By refining the clean-CVD and annealng conditions, further improvements in SOI width and crystallinity will be realized.

SPE of amorphous silicon is now being thoroughly investigatd and is expected to provide an important key in revealing crystal growth mechanisms and new LSI processes in the near future.

Acknowledgements

The authors are sincerely grateful to Mr. T. Sakai for his advice and encouragement throughout this work, and also to Mr. J. Nakata for his high energy ion beam experiments.

REFERENCES

1) Y. Kunii, M. Tabe, and K. Kajiyama: Jpn. J. Appl. Phys. **22** (1983) Suppl. 22–1.
2) Y. Kunii, M. Tabe, and K. Kajiyama: J. Appl. Phys. **54** (1983) 2847.
3) Y. Kunii, M. Tabe, and K. Kajiyama: Jpn. J. Appl. Phys. **21** (1982) 1431.
4) M. W. Jenkins: J. Electrochem. Soc. **124** (1977) 757.
5) K. E. Bean and P. S. Gleim: J. Electrochem. Soc. **110** (1963) 265C.
6) W. H. Shepherd: J. Electrochem. Soc. **112** (1965) 988.
7) M. Taniguchi, M. Hirose, and Y. Osaka: J. Crystal. Growth **45** (1978) 126.
8) D. C. Booth, D. D. Allred, and B. O. Seraphin: J. Non-Cryst.Solids **35 & 6** (1980) 213.
9) N. Sol, D. Kaplan, D. Dieumegard, and D. Dubreuil: J. Non-Cryst. Solids **35 & 6** (1980) 291.
10) M. Hirose, M. Taniguchi, T. Nakashita, Y. Osaka, T. Suzuki, S. Hasegawa, and T. Shimizu: J. Non-Cryst. Solids **35 & 6** (1980) 297.
11) Y. Ban, H. Tsuchikawa, and K. Maeda: Semiconductor Silicon (Abstract of Electrochem. Soc. Meeting) (1973) p. 292.
12) L. Csepregi, E. F. Kennedy, and J. W. Mayer: J. Appl. Phys. **49** (1978) 3906.
13) 2 G. Foti, J. C. Bean, J. M. Poate, and C. W. Magee: Appl. Phys. Lett. **36** (1980) 840.
14) 2 A. Lietoila, A. Wakita, T. W. Sigmon, and J. F. Gibbons: J. Appl. Phys. **53** (1982) 4399.
15) 2 R. Drosd and J. Washburn: J. Appl. Phys. **53** (1982) 397.
16) 2 M. Tabe, K. Arai, and H. Nakamura: Surf. Sci. **99** (1980) L403.
17) 2 K. Zellama, P. Germain, S. Squelard, J. C. Bourgoin, and P. A. Thomas: J. Appl. Phys. **50** (1979) 6995.
18) L. Csepregi, E. F. Kennedy, T. J. Gallagher, and J. W. Mayer: J. Appl. Phys. **48** (1977) 4241.
19) J. Nakata and K. Kajiyama: Appl. Phys. Lett. **40** (1982) 686.

Silicon-on-Insulator: Its Technology and Applications, edited by S. Furukawa, pp. 231–248.
© KTK Scientific Publishers, Tokyo, 1985.

CHARACTERIZATION OF SOLID PHASE EPITAXIALLY GROWN Si FILMS ON SiO$_2$

M. TAMURA[1], M. MIYAO[1], T. TOKUYAMA[1], H. YAMAMOTO[2], H. ISHIWARA[2], and S. FURUKAWA[2]

[1]*Central Research Laboratory, Hitachi Ltd., Kokubunji, Tokyo 185, Japan*
[2]*Department of Applied Electronics, Tokyo Institute of Technology, Nagatsuda, Midoriku, Yokohama 227, Japan*

Abstract Lateral solid phase epitaxial (L-SPE) films of vacuum-deposited Si on SiO$_2$ have been examined mainly through TEM observation, and compared with the case for (100) vertical solid phase epitaxial (V-SPE) films. Typical defects in L-SPE layers having a rough growth front are high density twins and dislocations, the generation of which are independent of ⟨100⟩ and ⟨110⟩ laterally grown films. On the other hand, in V-SPE layers containing low density short dislocations, {111} facets are formed at the ⟨110⟩ directed oxide window edge. It is thought that this facet acts as a barrier for changing from V-SPE to L-SPE across the SiO$_2$ edge. Successive 1050°C high temperature heat treatment generates interstitial type faulted and unfaulted dislocation loops in Si$^+$ ion implanted V-SPE layers. This is in contrast to the reduced dislocation density resulting in L-SPE layers having no remaining amorphous regions after the first 600°C annealing.

1. Introduction

Amorphous layers on crystalline semiconductor substrates can be recrystallized by means of solid phase epitaxy. If we use vertical solid phase epitaxial (V-SPE) films as a seed, it is possible to laterally grow single crystal films on amorphous materials through solid phase, low temperature annealing at, for instance, 600°C.[1-3] Such lateral seeding epitaxy has been extensively investigated in the field of liquid phase epitaxy using radiation sources to realize SOI structures.

Only a limited number of papers, however, have so far reported on lateral solid phase epitaxy (L-SPE) of Si on insulating films.[1-3] In particular, crystal quality and morphology of L-SPE layers have not been investigated. In this paper, we report on TEM observation results of L-

SPE films, with emphasis on the effect of seeding pattern shapes, pattern orientation, Si$^+$ ion implantation and annealing temperature.

2. Experiments

The sample structure and parameters used for experiments are shown in Fig. 1. Insulating films were of thermally grown SiO₂ and were formed into various shapes, particularly stripes, circles and square patterns. For some samples, the so called LOCOS (Local Oxidatiin of Si) process was used to form insulating films. Amorphous Si (a-Si) films were deposited in an ultra high vacuum chamber. Silicon ion implantation into a-Si films was carried out for some samples, in order to get dense a-Si films. Detailed experimental conditions for sample preparation have been reported elsewhere.[3] The specimens were thinned in planar and cross section using both chemical and ion milling procedures. The thinned specimens were observed using Hitachi HS-700H electron microscope operated at 200 keV.

3. General Observations

When only conventional evaporation technique was used for forming a-Si films, L-SPE was not realized, even though single crystallized V-SPE was completed.[3] It has been postulated that this is mainly due to the internal stress produced by non-uniform shrinkage in volume of the porous a-Si films deposited on the SiO₂ patterns during annealing.[3] This internal stress in films enhanced the random nucleation rate in a-Si films on SiO₂, and had a strong influence on inhibition of the change from V-SPE to L-SPE.

This section will first provide an explanation of the fact that, in addition to the internal stress, facet formation in V-SPE layers at the oxide window edge also contributes to the difficulty of L-SPE growth. Figure 2, a TEM micrograph, shows that a-Si films deposited on square shaped SiO₂ patterns become polycrystallites after 600°C, 12 h annealing, although a-Si on Si substrates transforms into a single crystal with V-SPE growth, as was clearly understood from the inserted diffraction pattern. That is,

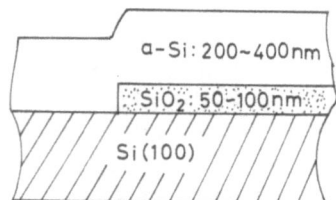

Fig. 1. Sample structure and parameters used for experiments.

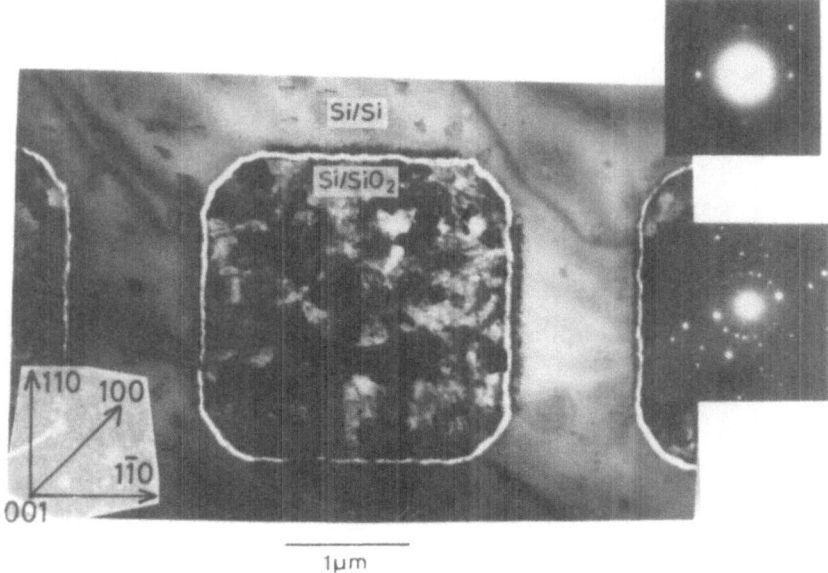

Fig. 2. TEM micrograph and TED patterns of a sample having square shape SiO₂ patterns for no L-SPE growth of deposited a-Si. 600°C, 12 h anneal.

L-SPE growth did not occur in this sample case. In the micrograph, it can be seen that two kinds of crystallographic defects are formed. One is a small defective zone formed at the SiO₂ edge in the ⟨110⟩ directions. Another involves short dislocations generated in single crystallized V-SPE regions. Dislocations showing a zig-zag contrast mainly run in the ⟨110⟩ directions and are inclined to the sample surface. From estimation of the almost constant length of these dislocations, it is judged that they originate from the interface between the deposited a-Si and the substrate.

Figure 3 is another example, and shows that lateral growth of deposited Si films did not occur from Si substrates into square shaped SiO₂ patterns fabricated using the LOCOS process. That is, a-Si films grew again into polycrystallites on SiO₂ patterns before the beginning of L-SPE growth. Moreover, in this sample, it is interesting to see that defective regions along the oxide window edge grow more expansively into single crystal areas than in the case represented in Fig. 2. This defect formation warrants further examination.

Figure 4, TEM micrographs were taken for defects generated at the oxide edge in the stripe oxide cut sample. Figure 4(a) is a planar TEM micrograph showing these defective zones, which are ⟨110⟩ elongated polycrystallites, apparently propagate from polycrystal regions into single

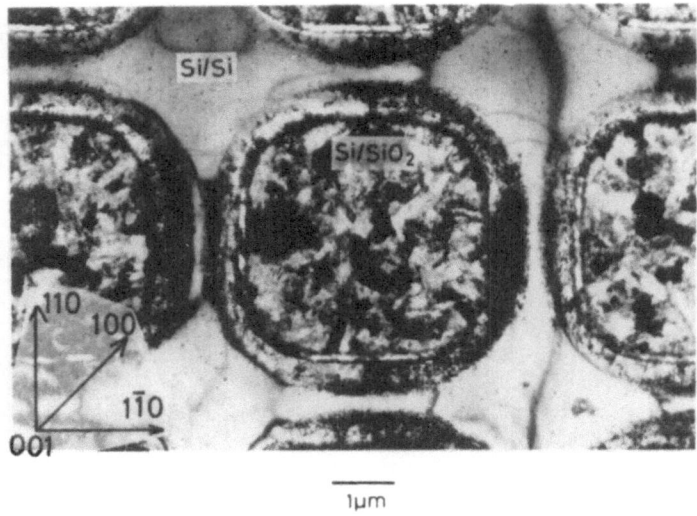

1μm

Fig. 3. TEM micrograph showing no L-SPE growth of a-Si deposited on square shape SiO₂ patterns made by LOCOS process. 600°C, 5h anneal.

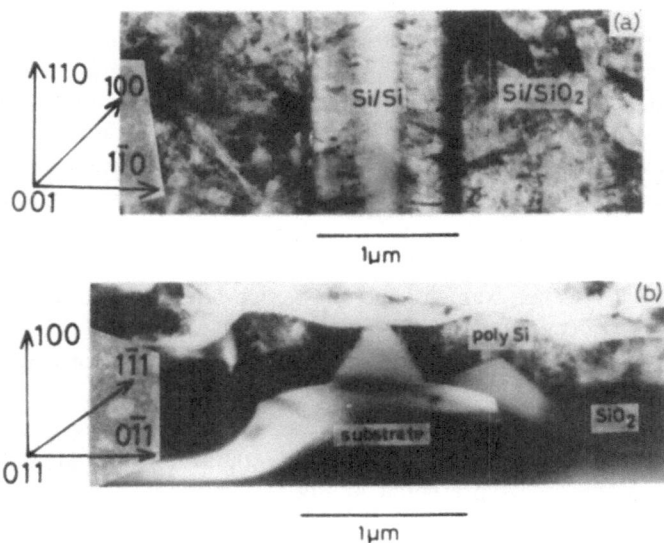

1μm

1μm

Fig. 4. TEM micrographs showing defects generated at the oxide edge in the stripe oxide cut sample. Sample structure and annealing conditions are the same as in Fig. 3 sample; (a) planar and (b) cross section TEM micrographs.

crystal areas, as was also shown in Fig. 3. Figure 4(b) is a (110) cross section TEM micrograph from a (100) specimen, corresponding to the Fig. 4(a) sample. From Fig. 4(b), we can distinctly recognize that facets with {111} planes, the slowest growing faces, are formed at the oxide window edge. This inclined facet formation results in a narrowing of single crystallized areas in V-SPE films, as can be seen in Figs. 3 and 4(a). In other words, defective zone propagation from the SiO₂ region to the substrate one is observed in planar TEM micrographs with poly-crystallite formation on facet planes.

Thus, it can be speculated that from the crystallographic point of view, facet formation at the SiO₂ edge acts as a barrier for change from V-SPE to L-SPE growth. Figure 5 shows the schematic growth sequence of {111} facets during V-SPE growth. Facets are considered to be already formed during the V-SPE growth process, as is shown in the figure. This suggests that the SiO₂ edge plays a pinning role in regard to the L-SPE growth of deposited films. That is, it can be easily anticipated that V-SPE films for seeding areas with facets hardly propagate across the SiO₂ boundary into the SiO₂ regions. Before lateral propagation by SPE growth into SiO₂ areas begins, polycrystallite formation occurs on the SiO₂ films. Such facet growth was also observed for a sample having a 50 nm thick thin SiO₂ film, as shown in Fig. 6.

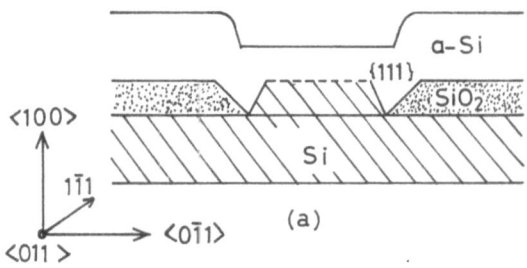

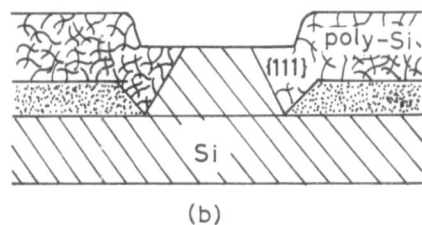

Fig. 5. Schematic growth sequence of {111} facets at the oxide edge during V-SPE growth; (a) during growth and (b) final stage.

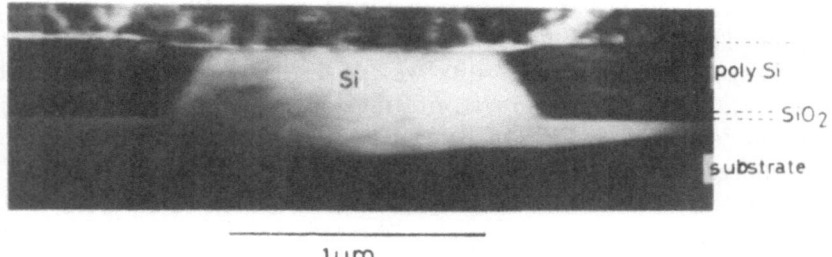

Fig. 6. Cross section TEM micrograph showing {111} facet formation in a V-SPE layer of a sample having 500 Å thick SiO₂.

Under optimum amorphous Si formation conditions, combined with Si^+ ion implantation[3], $\langle 110 \rangle$ directed L-SPE grew as shown in Fig. 7 in spite of the formation of {111} facets. That is, it can be considered that the {111} facet planes proceed laterally to the $\langle 110 \rangle$ directions on SiO_2 patterns after passing the SiO_2 edge barrier. From the planar micrograph in Fig. 7, we can see certain characteristics for grown L-SPE films.

One is a rough interface between amorphous and crystalline layers. This is in a strict contrast to the atomically smooth interface of V-SPE films between ion-implantation induced amorphous layers and crystalline substrates.[4] The rough interface in the present case may be due to the existence of some barriers when the growth mode changes from V-SPE to L-SPE. Namely, the beginning time for L-SPE growth is different for each position of the film along the SiO_2 edge. Furthermore, facets in the films may affect the lateral speeds for different areas of the film during growth. Another point is that L-SPE layers are very much defective, as observed from the micrograph showing $\langle 110 \rangle$ directed grain-like growth and diffraction patterns including twin spots. The defects (twins and dislocations) in the films are discussed in the next section.

In order to investigate in detail the relation between the orientation of films grown between V-SPE and L-SPE, micro-probe-RHEED measurements were performed. A field emission electron source with a probe beam focus of 0.1 μm was adapted.[5] The probe beam was, first, scanned over the whole sample to obtain a scanning electron microscope image on the display. Then, the beam was directed to the desired location on the sample by referring to the image. Thus, a diffraction pattern for a very small area (0.1 μm) could be obtained.

An absorption current image and diffraction patterns from various regions are shown in Fig. 8. The light field region for an absorption current image corresponds to a recrystallized layer. Single crystal diffraction

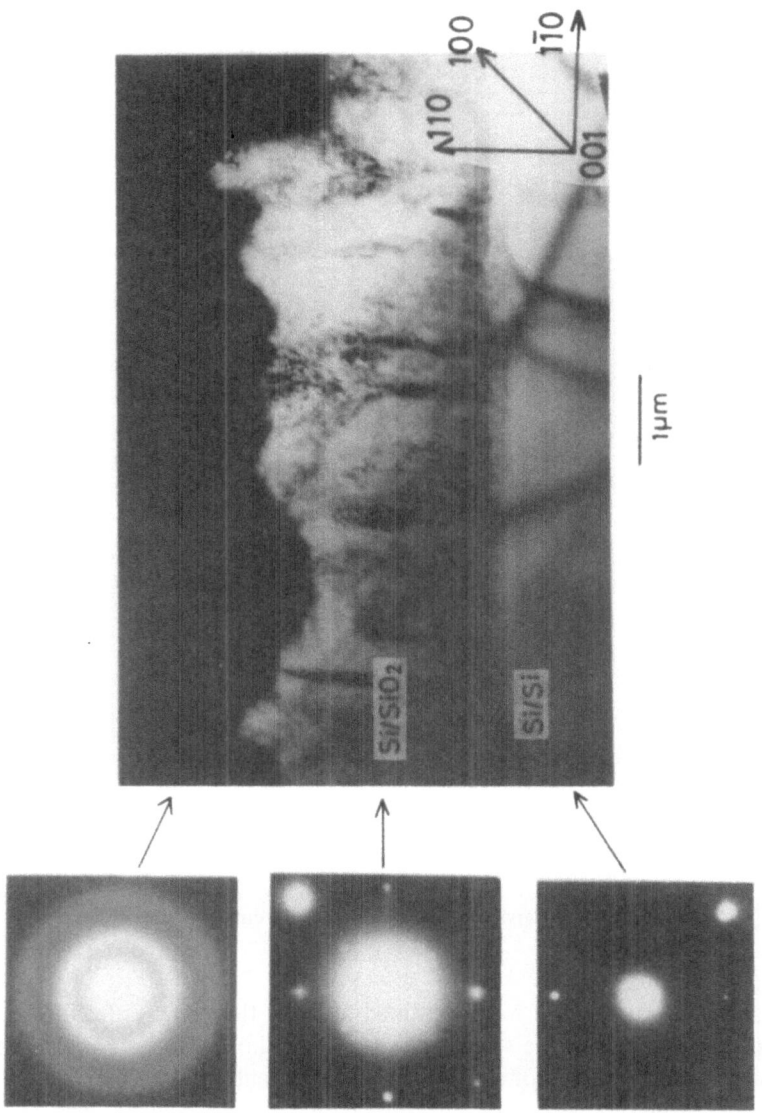

Fig. 7. TEM micrograph and TED patterns showing ⟨110⟩ directed L-SPE growth of a-Si. 600°C, 7h anneal.

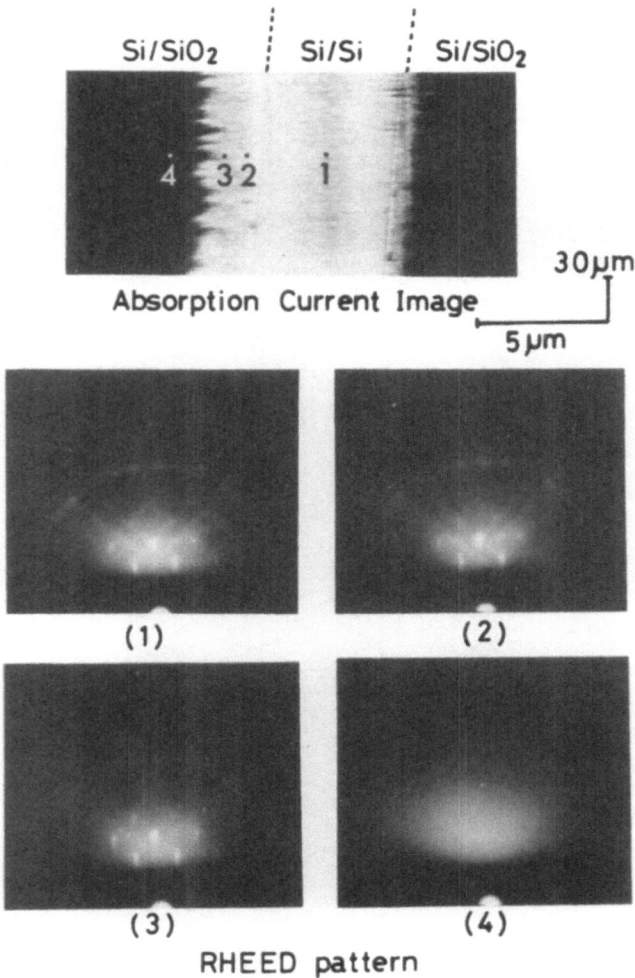

Fig. 8.   Micro-RHEED observation results showing seeding growth of L-SPE.

patterns with Kikuchi lines were obtained all over the recrystallized layers
((1) to (3)) as can clearly be seen in the figure, although diffraction intensities diminish gradually from (1) to (3). This indicates that crystal quality
degrades with the increase of the distance from the interface between Si/Si
and Si/SiO₂. Consequently, SOI formation by solid phase epitaxy as seeded from the substrate was precisely confirmed. On the other hand, diffraction pattern (4) from a dark field region indicates that this area still
remains in an amorphous state.

The lateral growth length of L-SPE in the ⟨100⟩ directions greater than that in the ⟨110⟩ directions in accordance with the results of the difference in V-SPE velocities in the ⟨100⟩ and ⟨110⟩ directions.[6] Figure 9 shows a TEM micrograph taken for a sample having oxide stripe windows patterned in ⟨100⟩ directions. It can be recognized from the figure that the lateral propagation length of the SPE reaches about 6 μm from the oxide window edge into the amorphous region. This length is roughly two times that of the 2-3 μm growth length for ⟨110⟩ directed L-SPE in Fig. 7. However, the growth front is still uneven, similar to the ⟨110⟩ pattern case. Moreover, the case shown in Fig. 9, formation of polycrystallites has already occurred in the remaining amorphous region. This suggests that lateral growth will not continue beyond this length.

4. Defects in Grown Films

Microtwins existed primarily in ⟨110⟩ grown L-SPE layers having {111} facet planes at the growth front. This is because twins are easily nucleated on {111} faces, as was pointed out during consideration of atom cluster matching on {111} faces with SPE growth.[7] Figure 10 is a (110) cross section TEM micrograph taken for the ⟨110⟩ laterally grown L-SPE film. This figure clearly shows that microtwins distribute on {111} faces mainly in the neighborhood of oxide edges. Moreover, we notice that other {111} facets different from those in V-SPE layers at the oxide edges (Figs. 4–6) are formed at the growth front of L-SPE layers. This situation is the same as for ⟨100⟩ laterally grown films with {110} facets as observed by SEM.[2]

However, microtwins also existed, even in ⟨100⟩ grown L-SPE layers with {110} facet planes. This may be because irregular lateral growth front generates {111} facet planes even during growth in ⟨100⟩ directed L-SPE layers. In Fig. 11, TEM micrographs taken near the growth front area of a ⟨100⟩ grown L-SPE layer are shown, clearly indicating the existence of ⟨110⟩ elongated high density microtwins. The upper dark field micrograph in the figure corresponds to the circled area in the bright field image. The diffraction pattern having twin spots was also taken from this region.

Another typical defect observed in L-SPE layers is dislocations. The TEM micrograph in Fig. 12 shows high density dislocation networks observed in a ⟨100⟩ grown L-SPE layer. As can clearly be seen from the micrograph, dislocations mainly run along the ⟨110⟩ directions. From this dislocation morphology, it can be speculated that dislocations do not originate from dislocation loop tangles which grow from implantation-induced point defect condensation.[7] Rather, dislocation morphology suggests that slipping on (100) planes occurs. A precise dislocation generation

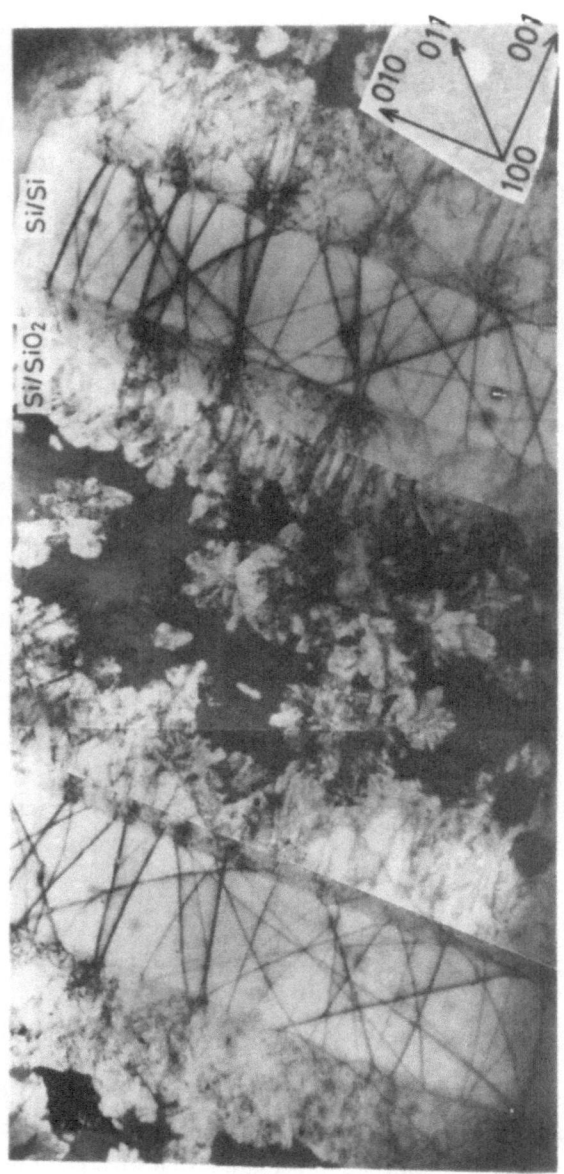

Fig. 9.   TEM micrograph showing ⟨100⟩ directed L-SPE growth of a-Si. 600°C, 8h anneal.

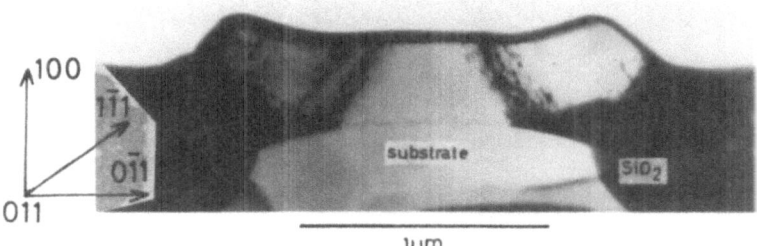

Fig. 10. Cross section TEM micrograph showing microtwin distribution in a ⟨110⟩ grown L-SPE layer.

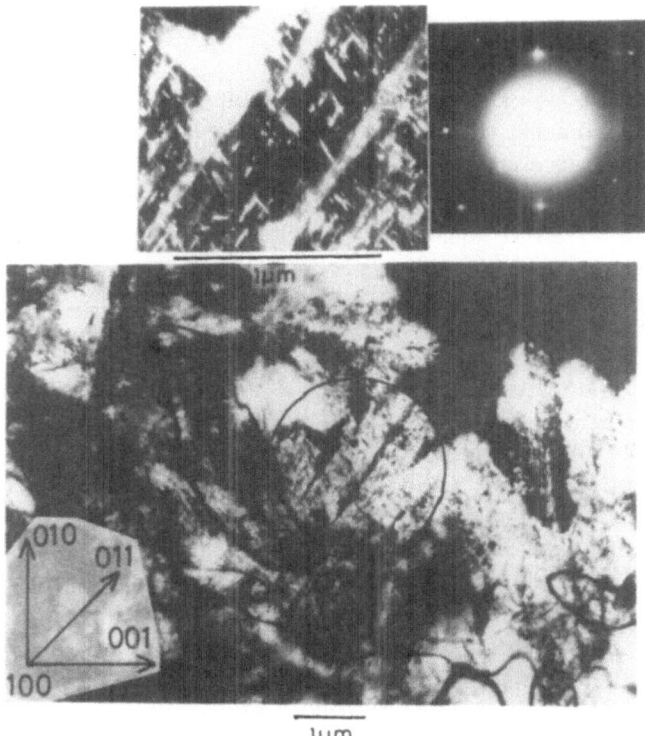

Fig. 11. TEM micrographs and TED pattern showing the existence of microtwins in a ⟨100⟩ grown L-SPE layer. An upper dark field image corresponds to the circled area in the bright field one.

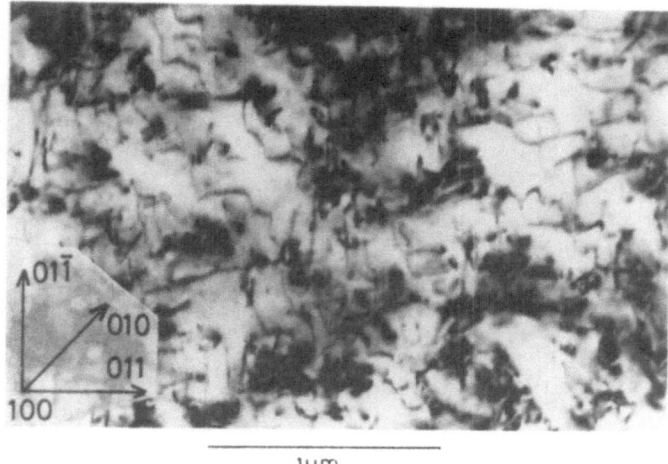

Fig. 12.   TEM micrograph showing high density dislocation networks in a ⟨100⟩ grown
L-SPE layer.

mechanism for L-SPE layers based on slipping has not been clarified at
the present moment. However, stresses induced by volume change from
an amorphous to a crystalline state during growth may be one cause for
dislocation generation. It can be also considered that mismatch stress at
the interface between crystallized Si and SiO_2 acts on the dislocation
formation.

5.   Effect of Seeding Pattern Shapes

This section describes the effect of seeding pattern shapes on the morphology of L-SPE growth. Figure 13 shows an example of a sample having repeated ⟨110⟩ directed stripe patterns in which all the regions, including both Si/Si and Si/SiO_2 areas, were crystallized even though L-
SPE regions are defective. We can note from the micrograph that an interaction phenomenon due to collision with the lateral growth from the
opposite window edges is hardly detectable near the central portion of
the films grown on the striped SiO_2 patterns. Only a weak boundary can
be slightly detected. This result contrasts markedly with the pulse laser
irradiated SOI formation results in which remarkably straight boundary
generation in the central region of grown films on the SiO_2 patterns resulted
from an intense collision of two opposite lateral growth movements.[8] The
difference between these lateral growth collision phenomena will be caused by the difference of the growth mode between solid and liquid phase
processes.

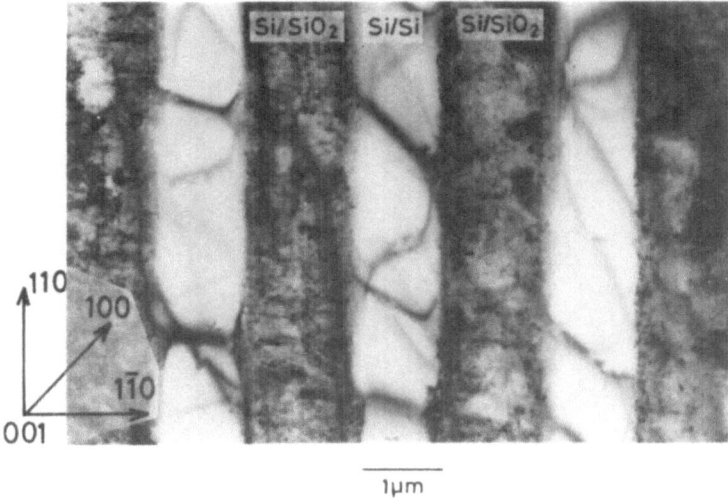

Fig. 13. TEM micrograph showing ⟨110⟩ L-SPE growth of a-Si on stripe SiO₂ patterns 600°C, 7h anneal.

The TEM micrograph in Fig. 14 was taken for a sample having circular oxide windows. In this case, all the areas in the sample were for the most part transformed into single crystals, excepting those hole areas (denoted by A in the figure) which still remain in an amorphous state. As far as can be speculated from the micrograph, leaving these A areas to be amorphous state, lateral epitaxy of a-Si on SiO₂ can be considered to have uniformly advanced from the seeds.

The sample in Fig. 15 has patterns reversed from those in Fig. 14. That is, the circular areas are Si/SiO₂ regions. The Si/SiO₂ areas were also single crystallized by the propagation of L-SPE growth from the surrounding seeds. However, no sign indicating collision of the growth front at the central portion in the circular areas and preferred oriented growth mode can be detected from the micrograph, just as with the results shown in Fig. 12.

6. *Effect of Successive High Temperature Annealing*

This section will cover the effect of successive high temperature annealing on 600°C grown SPE films. Figure 16 (a) is a typical example showing the effect of high temperature heat treatment, while Figure 16(b) is for a 600°C annealed sample. With high temperature annealing, the amorphous region remaining after 600°C annealing was transformed into

244 M. Tamura *et al.*

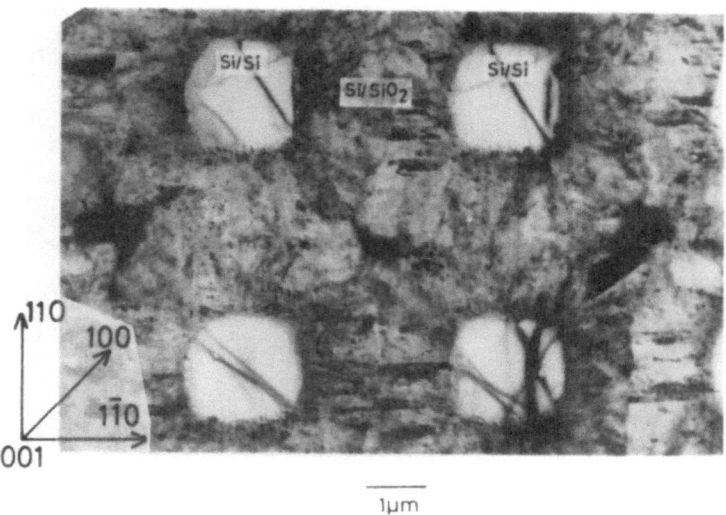

Fig. 14. TEM micrograph showing L-SPE growth of a-Si from circular oxide window areas. 600°C, 7h anneal. A in the figure denotes amorphous state.

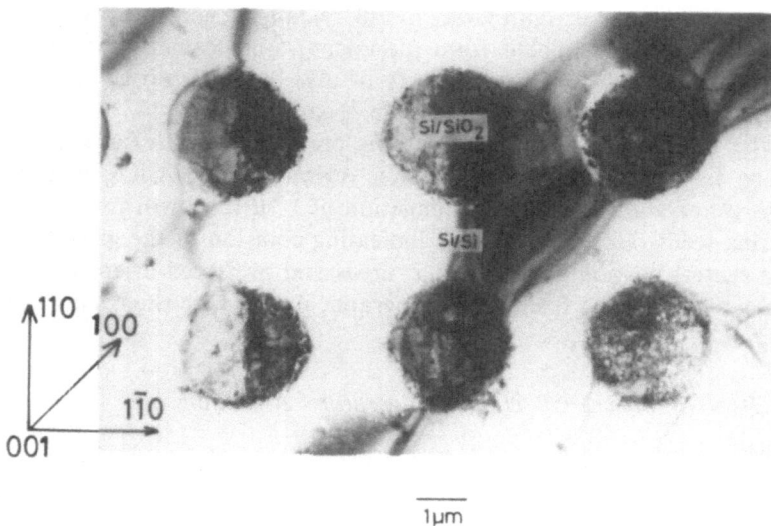

Fig. 15. TEM micrograph showing L-SPE growth of a-Si on circular SiO₂ patterns. 600°C, 7h anneal.

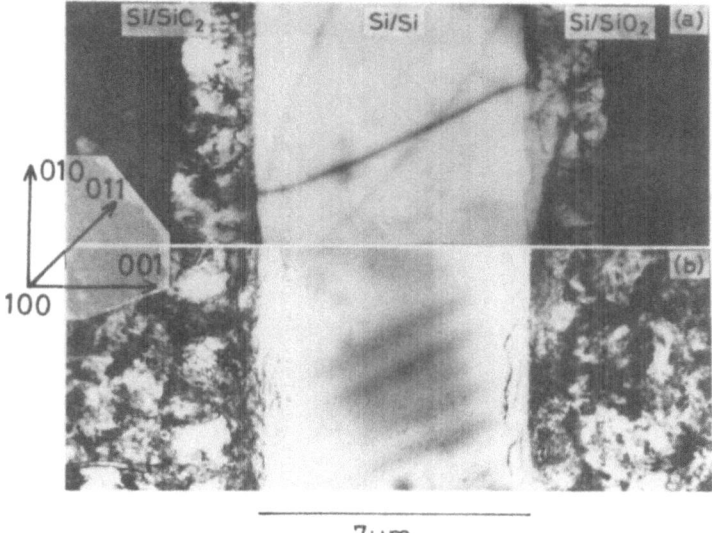

7μm

Fig. 16. TEM micrographs showing the effect of 1050°C heat treatment on the 600°C grown SPE films without Si⁺ ion implantation; (a) 600°C, 2h annealed sample, and (b) 600°C, 2h and 1050°C, 2h annealed sample.

polycrystals. It is not possible to distinguish from the micrograph the difference in crystal quality between previously formed L-SPE layers and polycrystal layers induced after annealing. Thus, further extension of L-SPE growth can hardly be expected with additional high temperature heat treatment. This is because of the drastic increase in random crystallization for a-Si films at high temperatures.[9]

On the other hand, defect generation in V-SPE layers other than some dislocation generation in the ⟨110⟩ directions in V-SPE layers within 1 μm from the oxide edges cannot be detected after high temperature annealing. In this case, Si⁺ ion implantation was not carried out in the initial a-Si films deposited on the sample.

However, dislocation loops grew in L-SPE layers of Si⁺ ion implanted and deposited a-Si films after high temperature annealing. The samples in Fig. 17 are ones with Si⁺ ion implanted amorphous Si films. In this figure, a TEM micrograph taken for the sample after 600°C annealing is compared with a micrograph of the sample following 1050°C annealing subsequent to the 600°C annealed one. As can clearly be seen, a great number of faulted and unfaulted dislocation loops are generated in V-SPE layers after high temperature annealing. The loops are 200–300 nm in diameter and mainly lie on {111} planes. In contrast to this, no disloca-

246   M. Tamura *et al.*

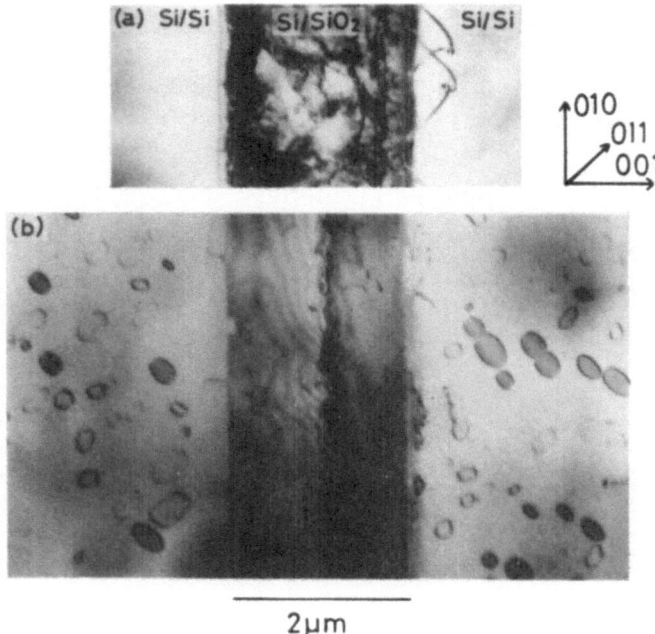

2μm

Fig. 17.   TEM micrographs showing the effect of 1050°C heat treatment on the 600°C grown SPE films with Si$^+$ ion implantation (180 keV, 2×10^{15} ions/cm^2 + 70 kev, 2×10^{15} ions/cm^2); (a) 600°C, 8h annealed sample, and (b) 600°C, 8h and 1050°C, 2h annealed sample.

tion loop generation occurs in the L-SPE layers. Instead, only a small, reduced number of randomly oriented dislocations remain, and a group of dislocation tangles run straight along the center of the L-SPE layers on the SiO$_2$ stripe pattern.

Formation of the loops in the V-SPE layers may be due to the coagulation of implanted excess Si interstitials or vacancies after high temperature heat treatment. On the other hand, there is no loop formation in the L-SPE layers because a number of sinks exist for the Si interstitials or vacancies in the defective L-SPE layers. The dramatic decrease in dislocation density of L-SPE layers due to high temperature annealing in Fig. 17 results in a strong contrast with the degradation result for L-SPE crystals shown in Fig. 16. This difference is due to whether the amorphous regions remain in the sample after the first 600°C annealing or not.

The dislocation loop type was determined from a contrast analysis experiment. The procedure for it was based on the formal method described by Hirsh *et al.*[10] In this method, the Burgers vectors, *b*, for the loops are defined by the right-hand convention. The image contrast for the loops changes from outside to inside for the real position when the quantity, $(g \cdot b)s$, changes from positive to negative. Here, s is a measure of the deviation of reflection vector g from the reflecting sphere. The shift in the dislocation image and change in loop size for bright-field images were carefully noted upon tilting the specimen from $+g$ to $-g$ (at $s>0$). Analysis showed that they were predominantly interstitial in nature.

7. Summary

Lateral solid phase epitaxy (L-SPE) of Si on SiO₂ was realized subsequent to completion of vertical solid phase epitaxy (V-SPE) of Si on Si. TEM observation results for the L-SPE were compared with V-SPE observation results. The main results are as follows.

With 600°C annealing:

(1) In samples having ⟨110⟩ SiO₂ patterns, L-SPE growth was often inhibited by growth of {111} facets at the window edge.

(2) Preferential ⟨110⟩ directed grain-like growth occurred in L-SPE layers. However, interaction phenomenon resulting from collision with growth from opposite window edges was not necessarily detected near the central portion of grown films on striped SiO₂ patterns.

(3) Typical defects in L-SPE layers were both mainly ⟨110⟩ oriented, high density dislocations and twins. On the other hand, only short, low-density dislocations existed in V-SPE layers.

With 600°C + 1050°C annealing:

(1) Interstitial type faulted and unfaulted dislocation loops (200–300 nm in diameter) grew in Si^+ ion implanted V-SPE layers. In contrast to this result, only randomly oriented dislocations remained in Si^+ implanted L-SPE layers, mainly along the central portion of grown films on the SiO₂ pattern.

(2) No extension of L-SPE was observed, as opposed to the results for 600°C annealing only. Rather, already formed L-SPE was degraded due to random crystallization of the remaining a-Si films.

REFERENCES

1) Y. Ohmura, Y. Matsushita, and M. Kashiwagi: Jpn. J. Appl. Phys. **21** (1983) L152.
2) Y. Kunii, M. Tabe, and K. Kajiyama: Jpn. J. Appl. Phys. (1983) Suppl. 22–1, 605.
3) H. Yamamoto, H. Ishiwara, S. Furukawa, M. Tamura, and T. Tokuyama: this volume.
4) J. Narayan: J. Appl. Phys. **53** (1982) 8607.
5) M. Ichikawa and K. Hayakawa: Jpn. J. Appl. Phys. **21** (1982) 145.
6) G. L. Olson, S. A. Kokorowski, J. A. Roth, and L. D. Hess: *Proc. Symp. Laser-Solid Interactions and Transient Thermal Processing of Materials, Boston, 1982*, ed. J. Narayan, W. L. Brown, and R. A. Lemons (North-Holland, New York, 1983).
7) R. Drosd and J. Washburn: J. Appl. Phys. **53** (1982) 397.

8) M. Tamura, M. Ohkura, and T. Tokuyama: Jpn. J. Appl. Phys. (1982) Suppl. 21–1, 193.
9) J. A. Roth, S. A. Kokorowski, G. L. Olson, and L. D. Hess: *Proc. Symp. Laser and Electron-Beam Interactions with Solids, Boston, 1981*, ed. B. A. Appleton and G. K. Celler (North-Holland, New York, 1982) p. 169.
10) P. B. Hirsh, A. Howie, R. B. Nicholson, D. W. Pashley, and M. J. Whelan: *Electron Microscopy of Thin Crystals* (Butterworths, London, 1965) p. 263.

CHAPTER 4 : CHARACTERIZATION AND DEVICE
APPLICATIONS

Silicon-on-Insulator: Its Technology and Applications, edited by S. Furukawa, pp. 251–261
© KTK Scientific Publishers, Tokyo, 1985.

MICROSTRUCTURAL CHARACTERIZATION OF SILICON-ON-INSULATOR STRUCTURES

R. F. PINIZZOTTO*

Central Research Laboratories, Texas Instruments Incorporated, P.O. Box 225936, M.S. 147 Dallas, Texas 75265, U. S. A.

Abstract Laser recrystallized silicon-on-oxide, graphite strip heater recrystallized silicon-on-oxide and buried oxide by high dose oxygen ion implantation are three of the silicon-on-insulator technologies currently being evaluated for VLSI and VHSIC applications. This report compares the microstructures of these materials.

The main defects in both the laser and graphite strip heater recrystallized material are subgrain boundaries. The misorientations across the boundaries are normally less than one degree. The boundaries are formed by dislocation coalesence. The dislocations are generated by the stresses caused by volume expansion during solidification of silicon droplets trapped in the solid silicon matrix. The droplets are caused by constitutional supercooling.

Buried oxide SOI formed by high dose oxygen ion implantation has fewer crystallographic defects than the other two materials. The entire ion implanted area is a single crystal after high temperature annealing. The primary defects in epitaxial layers grown on implanted substrates are dislocations. The oxide/silicon interfaces are abrupt if the dose is large enough. The sharp interfaces are formed by an internal oxidation mechanism. The minimum MeV He ion channeling yield from epitaxial silicon layers grown on buried oxide is about 2.5%, the lowest of any SOI material.

1. Introduction

Silicon-on-insulator (SOI) structures are being studied for use in very large scale and very high speed integrated circuits (VLSI and VHSIC) to exploit the advantages these materials have over bulk single crystal silicon. Soft errors caused by transient radiation can be minimized because the volume for electron-hole pair generation is reduced. CMOS latch-up and parasitic capacitance can be reduced by electrically isolating neighboring devices from one another by the formation of device islands. This permits

---

*Present address: Ultrastructure, Inc., 1850N. Greenville Ave.–140, Richardson, Tx. 75081, U.S.A.

a larger packing density since device cross-talk is negligible. The circuits have better high voltage isolation. There is a ground plane which may be easily accessed by cutting vias through the insulator. Some of the SOI technologies can be applied to three dimensional circuit integration. If a method of recrystallization is developed to form one device quality silicon layer, repetitive processing may result in an entire series of silicon layers separated by insulating layers. This has already been demonstrated by Irita et al. using high dose oxygen ion implantation and epitaxial layer deposition to form a triple layer SOI system. [1]

For the reasons cited above, many research groups are developing SOI fabrication methods. [2] It is absolutely necessary to understand the microstructures of these materials in order to improve both their crystal quality and to understand some of the variations in device performance as a function of material type. Since it is not known which type of SOI is best for VLSI and VHSIC applications, the various materials must be compared. This paper will review the current status of laser and graphite strip heater recrystallized, and buried oxide SOI materials.

2. Laser Recrystallized Laterally Seeded Silicon-on-Oxide

Lateral seeding was used to propagate single crystal silicon over an oxide region. The technique is based on forming single crystal over a seed and moving the growth front horizontally until the crystal extends over the oxide region. [3-6]

The misorientation both in the plane of the film and perpendicular to the plane of the film is normally $<4°$. The boundaries are composed of dislocation arrays. This is illustrated in Fig. 1, a montage of cross sectional TEM micrographs. The insert is a high magnification weak beam image of two of the subgrain boundaries. As expected, the misorientations are greater for those boundaries with larger dislocation densities. In the figure, the misorientation from a to b is 1.08 degrees, whereas that from b to c is 0.33 degrees. The changes in contrast in the top silicon layer are due to the small changes in orientation from subgrain to subgrain. The area shown in Fig. 1 is near the trailing edge of one of the pads. The lateral seeding process has broken down and the material is no longer a single crystal.

Because of the large difference in the thermal conductivity of silicon and silicon dioxide, [7] it is not possible to optimize the laser recrystallization process on both materials simultaneously. The silicon on silicon regions require more laser power since the silicon substrate absorbs a large fraction of the heat generated by absorption of the laser If a laser power sufficient to grow defect free single crystal in the seed regions is used, the silicon on top of the oxide pads absorbs far too much power and

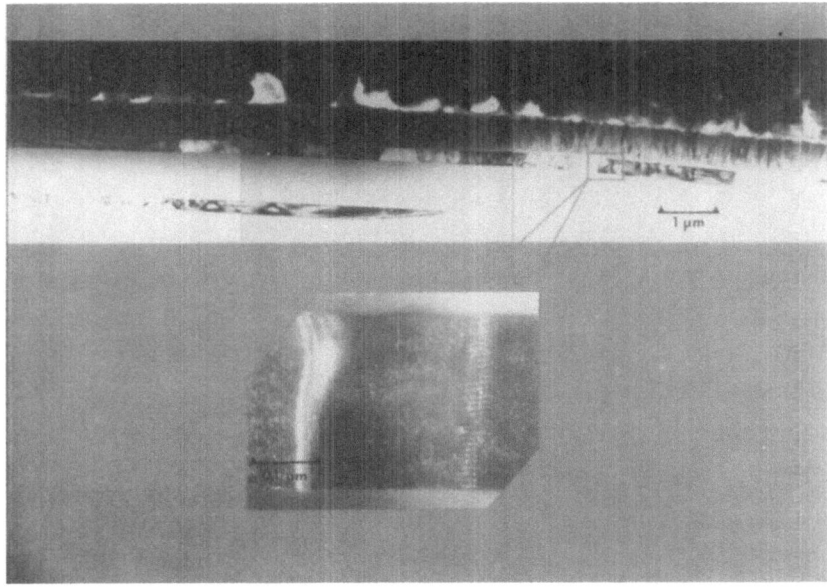

Fig. 1. Cross sectional TEM montage of l-lasso SOI. The insert is a weak beam image showing that the boundaries are composed of dislocation arrays.

is actually vaporized or "blown off" the substrate. Therefore, the laser power is adjusted to maximize lateral overgrowth and not to minimize the number of defects in the seeds. This effect is dramatically illustrated in Fig. 2, an HVEM micrograph of the defects in a typical seed region of l-lasso material. The primary defects are dislocations. The dislocations generated in the seed regions can propagate into the silicon-on-oxide areas. If the laser power is increased so that stacking faults or microtwins are formed in the seeds, they also may extend into the silicon-on-oxide regions.

3.   Graphite Strip Heater Recrystallized Laterally Seeded Silicon-on-Oxide

The primary defects in g-lasso material are subgrain boundaries similar to those found in the laser recrystallized samples. [8-11] Figures 3a and 3b are NDIC micrographs and 3c and 3d are TEM micrographs. The linear defects are subgrain boundaries. It is obvious from Fig. 3d that these boundaries are also composed of dislocation arrays.

Figure 4 shows two subgrain boundary initiation sites in g-lasso material. Most of the material is defect free single crystal silicon. However, a boundary will occassionally form in the middle of a defect free region.

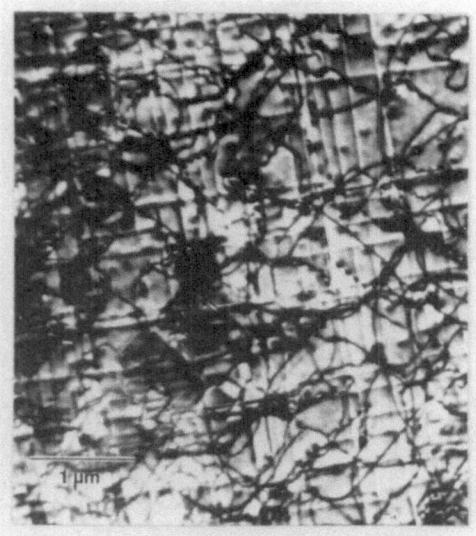

Fig. 2. HVEM micrograph of the defects in a typical l-lasso seed.

Fig. 3. A series of micrographs showing that the primary defects in g-lasso material are subgrain boundaries. The boundaries are composed of dislocation arrays.

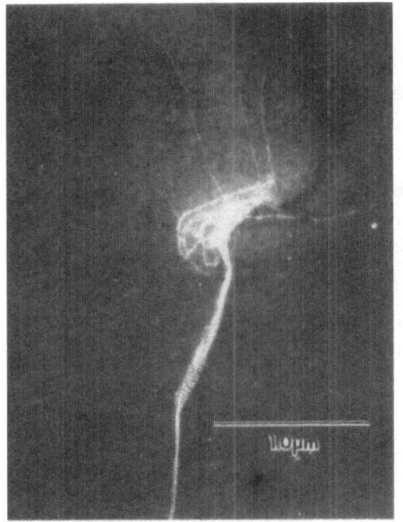

Fig. 4. Subgrain boundary initiation sites in graphite strip heater recrystallized lasso.

The initiation site is approximately circular with a very large dislocation density. A subgrain boundary propagates from this cluster in the direction of the top graphite strip heater motion. There are four major causes of dislocation formation in crystal growth: particulate inclusions, growth front impingement, non-uniform heat flow and point defect condensation. Particulate inclusions have been eliminated as a cause of boundary formation by careful control of the processing conditions. Point defect condensation can be ruled out because the defect morphology cannot be explained by this method. There are no isolated dislocation loops or stacking faults, as would be expected with an active point defect condensation mechanism in silicon. Non-uniform heat flow, while obviously present in these materials, cannot generate large enough stresses to homogeneously generate dislocation loops. [12] Growth front impingement of some sort must therefore be the root cause of subgrain boundary formation. Non-uniform growth of the single crystal can lead to liquid entrapment. When the liquid silicon freezes, it must expand by 9%. [7] This large change in volume can only be accommodated by stresses in the material, point defect generation and dislocation formation. The dislocations coalesce and form stable subgrain boundaries. Once the subgrain boundaries are formed, they propagate by normal dislocation assisted crystal growth. The existence of non-uniform solidification of silicon when radiatively heated is now firmly established. Chikawa and Sato first observed liquid inclusions in silicon using X-ray topography. [13] Bosch and Lemons have also shown, using videotape recording of the growth front during solidification, that long streamers and drops can exist up to several tens of microns from the advancing growth front. [11,14] Lemons has recently proposed that the interface structure is caused by consititutional supercooling. [14] Tiller et al.'s model [15] predicts that substructure should not form below a critical value of the temperature gradient divided by the growth velocity. Lemons was not able to demonstrate this effect. Celler et al. have been able to grow subgrain boundary free material by recrystallizing thick polysilicon layers (< 15 um) over entire wafers with tungsten lamps. [16] It is not known at the present time whether the boundaries were eliminated because of the low recrystallization velocities or due to an inherent property of the thick polysilicon layer itself.

4. SOI Formed by High Dose Oxygen Ion Implantation

Ions implanted with large kinetic energies are deposited below the surface with an approximately Gaussian depth distribution. Near the surface, the concentration of the implanted ion can be quite small. However, the crystal damage versus depth distribution is completely different. The position of maximum damage occurs slightly closer to the sample surface than the position of maximum implanted ion concentration. There is a critical value of the damage level above which the target becomes either polycrystalline or amorphous. If the surface damage level remains below this critical value, the surface layer is still a single crystal after the implantation. This can be achieved by heating the wafer during implantation, either externally or by the ion beam itself. After implantation, but before high temperature annealing, there is a thin single crystal (100) surface layer on top of an amorphous layer. There is a layer of silicon dioxide (if oxygen ions are implanted), another damage layer and finally the (100) silicon substrate. After annealing, the entire region above the buried layer is converted to (100) single crystal silicon by solid state epitaxial regrowth. The oxide is stoichiometric SiO_2.

In a sample implanted with a dose of 1.46×10^{18} O cm^{-2}, a distinct buried oxide region is formed. There is a layer of polysilicon on both sides of the oxide layer. A similar structure was recently reported by Hayashi et al. [17] As the dose is further increased, the polycrystalline layers are oxidized until the oxide to silicon interfaces on both sides of the buried oxide layer are abrupt. The first report of abrupt interfaces was based on sputter AES data, again by the Japanese group headed by Hayashi. [18] During the early stages of the implant, the oxygen ions come to rest in a nearly Gaussian distribution inside the silicon lattice. As the dose is increased further, the oxygen concentration becomes too large for the silicon to retain its crystallinity and the lattice is broken down. The buried layer is a mixture of polysilicon plus oxide. As the implant continues, the concentration eventually becomes high enough near the projected range to form stoichiometric SiO_2. A distinct buried oxide layer is formed. With increasing amounts of oxygen, the oxide layer grows by consuming more and more of the polysilicon on either side of it. This is possible because the oxide remains stoichimetric, that is the local ratio of oxygen to silicon is fixed at 2. During the implantation, the oxygen comes to rest at the projected range, to a first approximation. The oxide has a high defect density and the diffusivity of oxygen in the oxide is much larger than in normal thermal annealing. The oxygen can rapidly diffuse to the interfaces where it reacts and forms SiO_2. This process continues as the dose is increased until eventually the diffusion length of the oxygen is significantly larger than the distribution width of the implanted oxygen. At this point, abrupt oxide/silicon interfaces are formed.

The standard post-implantation anneal sued by most research groups is two hours at 1150°C.

Typical results from Rutherford backscattering spectroscopy and MeV He ion channeling experiments are shown in Fig. 5. The random spectrum of the implanted sample shows the position of the buried oxide layer by

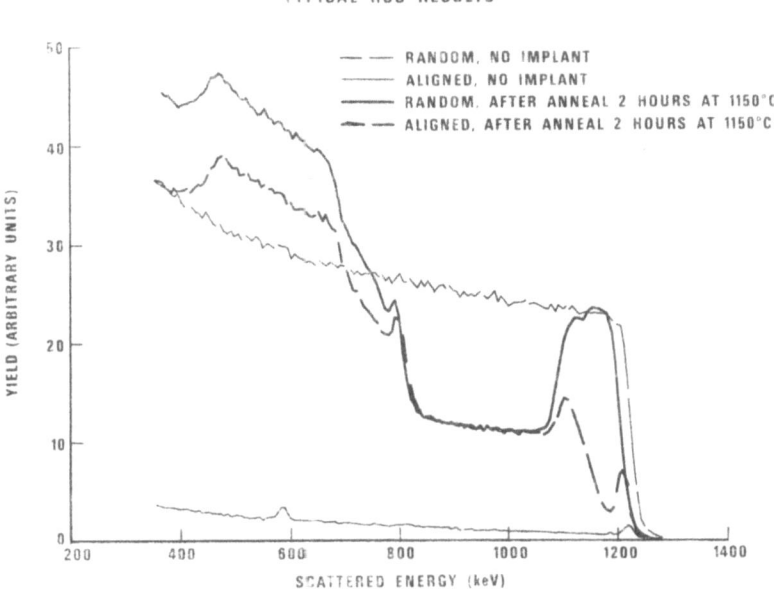

Fig. 5. Rutherford backscattering spectra for unprocessed single crystal silicon and buried oxide SOI both randomly aligned and channeled along the ⟨110⟩ direction.

both a decrease in the backscattered intensity due to a lower density of silicon in the oxide layer and an increase in intensity at the low energy end of the spectrum due to the presence of oxygen. From Fig. 5, the thickness of the top silicon layer is approximately 170 nm and the buried oxide is 500 nm thick. This is in excellent agreement with measurements by TEM and SEM cross sections and estimated values from sputter AES analysis. It should be noted the small peaks near 800 keV in both spectra are due to a very thin layer of oxide on the surface of the sample after annealing.

In Fig. 5, the minimum backscattering yield of an unimplanted sample channeled along the ⟨220⟩ direction is shown. There is a small surface peak, which is a characteristic of the technique, due to small misalignments and reconstructions at the surface. The small peak near 580 keV is due to a thin carbon layer on the surface of the sample. This can arise either from contamination of the sample by the atmosphere or by contamination caused by the pumping system of the RBS apparatus. The aligned channeled spectrum for an implanted and annealed sample is also shown. There is a much larger surface peak followed by a sharp decrease in the yield. This is indicative of epitaxially regrown material in the top silicon layer. The sharp increase in yield as a function of depth (lower energy) is due to defects that are present in the top layer near the silicon/oxide interface. All the ions are dechanneled by the buried oxide layer, hence there is no channeling information from depths greater than the top layer thickness. The minimum backscattered channeling yield was monitored as a function of anneal time in both Ar and N_2 ambients. The value of the minimum yield is a good qualitative measure of the material in the top layer. This data is plotted in Fig. 6. The minimum yield decreases as a function of time for both ambients. For N_2 annealing, the equilibrium value is near 35% and is reached in approximately 200 minutes. The results obtained with Ar ambients are superior to those with N_2 ambients. The decrease in backscattering yield is more rapid and the final value obtained was lower. The minimum yield approaches 20% after 240 minutes in Ar at 1150°C.

NDIC observations after annealing showed that the surfaces of samples annealed in N_2 became pitted. Both the pit density and size are functions of anneal time. The pit densities for samples annealed in Ar were several orders of magnitude smaller than those annealed in N_2. Homma et al. also observed this effect and found that the pits extended to the top silicon layer/buried oxide interface. [19] Since the ion beam used for RBS analysis is much larger than the pit size, both the top silicon layer and the buried oxide uncovered by the pits will be analyzed simultaneously. Channeling does not occur in the oxide, hence the minimum yield for pitted samples is larger than for intact top silicon layers. Cracks were found in samples

MINIMUM CHANNELING YIELD VERSUS ANNEAL TIME

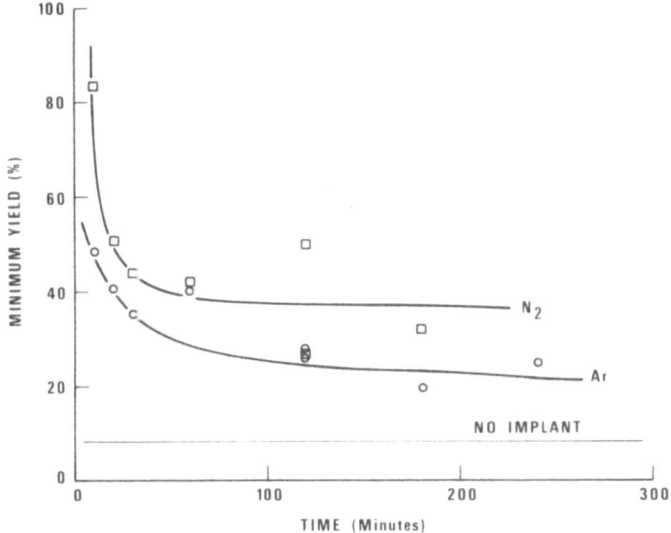

Fig. 6. The minimum channeling yield for buried oxide samples as a function of anneal time at 1150°C in N_2 or Ar.

annealed in both ambients. These may account for part of the difference in minimum yield between buried oxide SOI and unprocessed single crystal silicon.

Figure 7 shows RBS spectra of a buried oxide sample after deposition of an 0.4 um thick epitaxial silicon layer, both randomly aligned and channeled along the ⟨220⟩ direction. The minimum yield is 2.5%, the same value obtained for unprocessed wafers. This is the best value obtained for any SOI material. The channeled yield remains small for almost the entire epitaxial layer thickness, indicating that the defect density is low until the oxide/substrate interface is reached.

5. Summary

Laser recrystallized laterally seeded silicon-on-oxide consists of large grained polysilicon after the growth surpasses approximately 100 μm. The grains are aligned to the [100] of the substrate to within 4 degrees. Graphite strip heater recrystallized laterally seeded silicon-on-oxide is "single crystal" over an entire 76 mm diameter wafer. The defects are small angle subgrain boundaries. The deviations across the boundaries are typically only a few

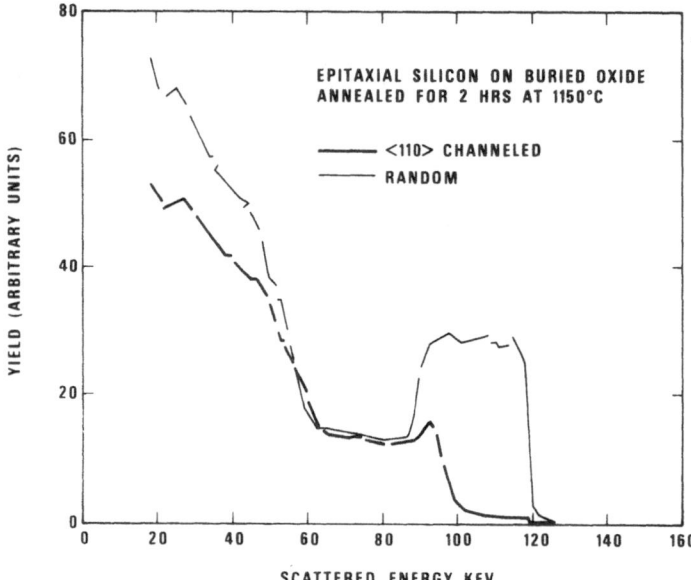

Fig. 7. Rutherford backscattering spectra from a buried oxide sample with an epi layer, both randomly aligned and channeled.

tenths of a degree. The orientation over large areas varies by at least 1 degree. Dislocations are the primary defects in both of these materials. The dislocations are formed by the stresses caused by the volume expansion of silicon during freezing. The SOI formed by high dose oxygen ion implantation is very high quality single crystal over the entire implanted area. The silicon/oxide interfaces are sharp and abrupt when large doses are used. The interfaces are formed by an internal oxidation mechanism. Epitaxial layers of very good quality may be grown on the material after suitable high temperature anneals.

Acknowledgements
    It is my privilege to acknowledge my co-workers for all of their efforts and discussions. H.W.Lam, S.D.S.Malhi and C.E.C.Chen are intimately involved with sample and device fabrication. B.L.Vaandrager worked tirelessly on the electron microscopy. T.J.Shaffner and R.A.Bowling helped with the Auger and ESCA spectroscopies. The RBS analyses were performed at NTSU by A.H.Hamdi and F.D.McDaniel. S.Matteson offered invaluable insight into the interpretation of the data. Much of the work was performed by a group of very talented technicians, especially M.L.Jarvis, who put up with working with us in the TEM lab every day, K.H.Madden, J.K.Russell, M.G.Nichols and J.A.Rushton. There were many, many others without whose efforts this project would not have been possi-

ble. The high voltage electron microscopy was performed at the HVEM-Tandem Facility of Argonne National Laboratory of the Department of Energy, A.Taylor, director. The work was supported by several contracts from the Office of Naval Research (M.Yoder and K.Davis), the Defense Advanced Research Projects Agency (S.Roosild), the Naval Ocean Systems Center (I.Lagnado)and the Rome Air Development Center (P.Vail, monitor).

REFERENCES

1) Y.Irita, Y.Kunii, M.Takahashi, and K.Kajiyama: Jpn. J. Appl. Phys. **20**(1982)L909.
2) The Extended Abstracts of the Electrochemical Society Spring Meeting, Montreal, Canada, May 1982, Vol. 82-1, The Electrochemical Society, Pennington, N.J., 1982. This volume contains the abstracts for the symposium, "Growth of Single Crystals on Amorphous Substrates."
3) H.W.Lam: *IEDM Tech. Digest* (IEEE, Dec. 1980) p. 556.
4) H.W.Lam, R.F.Pinizzotto, and A.F.Tasch, Jr.: J. Electrochem. Soc. **128**(1981)1981.
5) M.Tamura, H.Tamura, and T.Tokuyama: Jpn. J. Appl. Phys. **19**(1980)L23.
6) M.C.Flemings: *Solidification Processing* (McGraw-Hill, 1974).
7) H.F.Wolf: *Silicon Semiconductor Data* (Pergammon Press, Oxford, 1969).
8) R.F.Pinizzotto, H.W.Lam, and B.L.Vaandrager: Appl. Phys. Lett. **40**(1982)388.
9) E.W.Maby, M.W.Geis, Y.L.LeCoz, D.J.Silversmith, R.W.Mountain, and D.A.Antoniadis: IEEE Electron. Dev. Lett. **EDL-2**(1981)241.
10) M.W.Geis, H.I.Smith, B.Y.Tsaur, J.C.C.Fan, D.J.Silversmith, and R.W.Mountain: J. Electrochem. Soc. **129**(1982)2812.
11) H.J.Leamy, C.C.Chang, H.Baumgart, R.A.Lemons, and J.Cheng: Mat. Lett. **1**(1982)33.
12) W.C.Dash: J. Appl. Phys. **30**(1959)459.
13) J.Chikawa and F.Sato: Proc. Mater. Res. Soc. **2**(1981)317.
14) R.A.Lemons, M.A.Bosch, and D.Herbst: Proc. Mater. Res. Soc. **13**(1983)581.
15) W.A.Tiller, K.A.Jackson, J.W.Rutter, and B.Chalmers: Acta Metall. **1**(1953)428.
16) G.K.Celler, McD.Robinson, and D.J.Lischner: Appl. Phys. Lett. **42**(1983)99.
17) T.Hayashi, H.Okamoto, and Y.Homma: *Inst. Phys. Conf. Ser.*, No. 59, p. 533.
18) T.Hayashi, H.Okamoto, and Y.Homma: Jpn. J. Appl. Phys. **19**(1980)1005.
19) Y.Homma, M.Oshima, and T.Hayashi, Jpn. J. Appl. Phys. **21**(1982)890.

Silicon-on-Insulator: Its Technology and Applications, edited by S. Furukawa, pp. 263–268.
© KTK Scientific Publishers, Tokyo, 1985.

HIGH SPEED SOI-CMOS DEVICES BY LASER RECRYSTALLIZATION TECHNIQUE

T. Nishimura, Y. Akasaka, and H. Nakata

LSI Research and Development Laboratory, Mitsubishi Electric Corporation, 4-1, Mizuhara Itami 664, Japan

Abstract CMOS devices are fabricated on laser-recrystallized polysilicon islands on an insulating layer. Both n-channel and p-channel MOSFETs having channel length of 2 μm exhibit normal operation by controlling the grain boundary direction in the channel region. The low field electron and hole mobilities are 580 cm^2/V sec and 220 cm^2/V sec, respectively.

19-stage CMOS ring oscillators with nominal channel lengths of 3 μm are fabricated. The minimum propagation delay is 280 psec/stage at supply voltage of 10 V, and the minimum power delay product is 0.13 pJ/stage.

1. Introduction

Recently, silicon on insulating (SOI) films fabricated by various recrystallization methods have been significantly improved in crystal quality.[1-3] However, performance of MOS devices fabricated on them was still inferior to that of the devices on bulk silicon due to their longer channel lengths of 5 μm level and/or lower carrier mobility. This is mainly because grain boundaries which exsist on SOI after recrystallization cause fast diffusion of source and drain dopants, and prevent realizing the short channel devices.

We have reported [4] that a direction of grain boundary in laser-recrystallized polysilicon island could be arranged along with the laser scan direction by optimizing the laser irradiation condition. By using this technique, grain boundaries which act as fast diffusion paths of dopants in the channel region were almost eliminated.

In this study, CMOS devices were fabricated on the laser-recrystallized polysilicon islands. Both n-channel and p-channel MOSFETs having channel length of 2 μm were obtained with good uniformity and reproducibility. 19-stage CMOS ring oscillators with channel lengths of 3 μm were fabricated with a good functional yield. The speed performance was significantly improved to be 280 psec/gate, which is faster than any other reported value.[1-3]

2. Experiments

The starting material was a 4-inch single crystalline silicon wafer with 1.1 μm thick polysilicon by low-pressure chemical vapor deposition (LPCVD) at 610°C, the polysilicon film in the field region was converted to thermally grown oxide by selective oxidation. Device islands were surrounded by SiO_2 retaining walls and were completely isolated from each other. Polysilicon islands for p-channel and n-channel devices were implanted for substrate doping with phosphorus and boron, respectively. After removing any encapsulating layer on polysilicon islands, the laser recrystallization process was performed by using a cw-Ar laser for a whole 4-inch wafer. The laser power, scan speed, and scan step were 7.5 W, 12.5 cm/sec, and 20 μm/step, respectively, with a focused spot size on the sample of about 50 μm. Backside temperature of the sample was kept constant at 450°C. Laser beam was arranged to scan perpendicular to the channel direction of MOSFETs which should be fabricated in the island. The thickness of the gate oxide measured on the [100]-oriented reference wafer was 400 Å. Successive fabrication steps were carried out by a conventional CMOS-LSI process, in which source and drain regions of p-channel and n-channel MOSFETs were implanted with boron of $5 \times 10^{14}/cm^2$ and arsenic of $4 \times 10^{15}/cm^2$ by using photoresist masks, respectively. The feature of this CMOS process was single level polysilicon with two level aluminum metallization.

3. Results

Figure 1 shows threshold voltages of n-channel and p-channel MOSFETs with the boron dose of $3 \times 10^{12}/cm^2$ and phosphorus dose of $1 \times 10^{12}/cm^2$ as a function of the nominal channel length. Below 2.5 μm threshold voltages were decreased due to the short channel effect in both n-channel and p-channel cases. Lateral diffusion lengths of dopant impurities from both sides, the source and drain, into the channel region were deduced from the reciprocal plotts of current gain constants versus nominal channel lengths. Those for n-channel and p-channel MOSFETs were 0.7 μm and 0.8 μm, respectively.

Although the surface channel electron and hole mobilities obtained from the transconductance curves were decreased as the channel length increased due to the existence of grain boundaries perpendicular to the current flow, those for n-channel and p-channel MOSFETs increased to 580 cm²/V sec and 220 cm²/V sec at the nominal channel length of 2

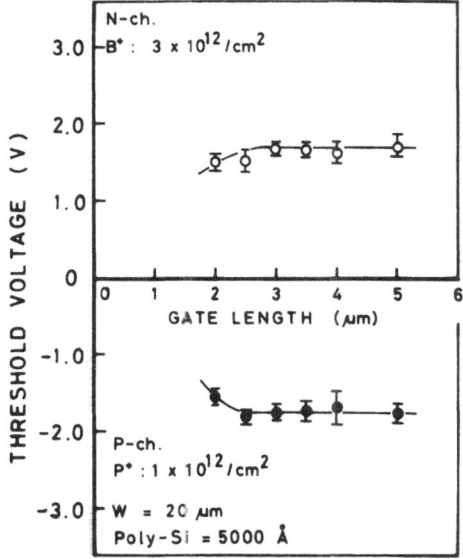

Fig. 1.   Threshold voltages of n-channel and p-channel MOSFETs as a function of channel lengths.

μm, respectively. These values were almost comparable to those of devices on bulk silicon.

Typical leakage currents for n-channel and p-channel MOSFETs were less than 0.5 pA/μm (current /unit channel width) for gate bias voltage of 0 V, and source to drain voltages of 5 V. The electrical characteristics stated above are summarized together with those on bulk silicon in Table 1.

19-stage CMOS ring oscillators were fabricated to assess the speed performance of the SOI structure and to achieve its application for functional circuit. The n-channel and the p-channel MOSFET with channel lengths of 3 μm were fabricated on individual device islands which were 20 μm long and 10 μm to 200 μm wide. The island width for the p-channel MOSFET was designed to be twice that of the n-channel MOSFET. Figure 2 shows the typical output waveform of the ring oscillator with the n-channel width of 10 μm. Performance of several oscillators with different threshold voltages, and that of the same device on bulk silicon are shown in Fig. 3. The minimum propagation delay/stage was 280 psec at supply voltage of 10 V with a power delay product of 2.2 pJ. The minimum power delay product was 0.13 pJ. The operating speed is faster than any other value on SOI reported so far, [1-4] and also quite the same as that of a device on bulk silicon due to the reduced parasitic capacitance.

The functional yield in a wafer exceeded 90% and the uniformity

Table 1. Summary of electrical characteristics of MOSFET (channel length; 3 μm, width; 20 μm).

| | SOI | | BULK Si | |
	N-ch.	P-ch.	N-ch.	P-ch.
Threshold voltage (V)	1.0	-1.6	0.65	-0.63
Source to drain break down voltage (V)	$12 \sim 13$	$-15 \sim -18$	$15 \sim 17$	$-18 \sim -20$
Carrier mobility (cm^2/Vsec)	580	200	600	230
Leakage current/ unit width (pA/μm)	0.5	0.5	0.05	0.05
Gate break down voltage (V)	$27 \sim 32$		35	

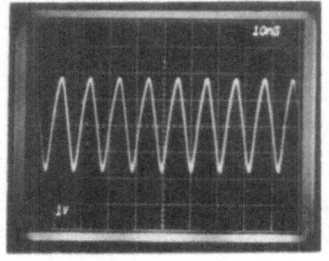

Vdd = 10 V

DELAY TIME
280 psec / stage

Fig. 2. Output waveform of 19-stage SOI-CMOS ring oscillator at $V_{DD} = 10$ V.

(a standard deviation/the average value) of the propagation delay time/stage at a supply voltage of 10 V was 4.6%.

4. Conclusion

SOI-CMOS devices were fabricated by using a laser-recrystallized

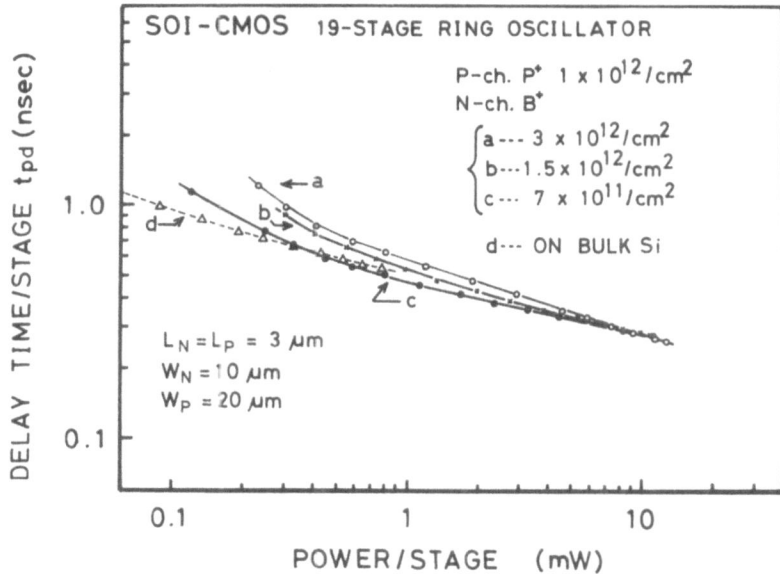

Fig. 3. Operating characteristics of 19-stage SOI-CMOS ring oscillators and the same ring oscillator on bulk silicon.

polysilicon islands on a whole 4-inch wafer. Since the source-to-drain short was eliminated by controlling the direction of grain boundary in the channel region to be perpendicular to the channel direction. This enables us to fabricate short channel devices of 2–3 μm with good uniformity and reproducibility.

CMOS ring oscillators with 3 μm channel length exhibited the minimum propagation delay/stage of 280 psec and power delay product of 0.13 pJ. This speed performance is superior to those reported so far and also to that of the same ring oscillator fabricated on bulk silicon. These results indicate that the laser-recrystallization technique is potentially applicable for achieving the SOI device of high speed and high packing density, and more complicated devices such as three dimensional circuits.

Acknowledgements
The authors are grateful to Dr. H. Oka for his interest and support of this research program. They are indebted to K. Sugahara and S. Kusunoki for help with experiments.

This work was performed under the management of the R&D Association for Future Electron Devices as a part of the R&D Project of Basic Technology for Future Industries sponsored by Agency of Industrial Science and Technology, MITI.

REFERENCES

1) S. D. S. Malhi, H. W. Lam, and R. F. Pinizzotto: IEDM 82, Extended Abstract, p. 441.
2) T. I. Kamins: Electron Dev. Lett. **EDL-3** (1982) 341.
3) B. Y. Tsaur, J. C. C. Fan, R. L. Chapman, M. W. Geis, D. J. Silversmith, and R. W. Mountain: Electron Dev. Lett. **EDL-3** (1982) 398.
4) T. Nishimura, A. Ishizu, and Y. Akasaka: Appl. Phys. Lett. **42** (1983) 102.

Silicon-on-Insulator: Its Technology and Applications, edited by S. Furukawa, pp. 269–281.
© KTK Scientific Publishers, Tokyo, 1985.

CHARACTERIZATION OF SOI DOUBLE Si ACTIVE LAYERS THROUGH FABRICATION OF ELEMENTARY DEVICES

M. MIYAO, M. OHKURA, and T. TOKUYAMA

Central Research Laboratory, Hitachi Ltd., Kokubunji, Tokyo 185, Japan

Abstract Various active regions in seeded lateral epitaxial Si layers obtained by cw scanning Ar laser irradiation are characterized by fabricating elementary devices. P-n diodes in the seeding area show a forward current with an ideality factor of 1.2–1.4 and a reverse current level of 6×10^{-8} A/cm^2 at 0.1 V. Electrical properties of MOSFETs located at the SiO$_2$ edge are somewhat poor ($\mu = 450$ cm^2/V.s., $V_{TH} = 1.0$V) due to the existence of dislocations and residual stress. However, the dislocations escape to the sample surface along the inclined plane of the SiO$_2$ edge during lateral growth. Thus, no crystal defects are observed in the Si layers grown over the SiO$_2$. The MOSFETs located on the front and back side of the SOI layer show high electron mobility (600 cm^2/V.s) and a reasonable threshold voltage (0.5V). A novel device structure, where the diode forward current is controlled by a MOS gate, is also presented. Unsaturated, triode-type characteristics are observed.

1. Introduction

Recent investigations of laser and electron beam annealing have focused on fabrication of Si on insulator (SOI) structures.[1-3] Grain growth in poly. Si on insulating layers,[4,5] bridging epitaxy,[6-8] i.e., seeded-lateral epitaxy,[9] and graphoepitaxy[10] are the major achievements to date. These techniques promise realization of new device structures, such as completely isolated, high speed devices and/or three-dimensional (3-D), high density LSI's.

Of these techniques, seeding epitaxy is regarded as the most realistic approach for VLSI application. This is because crystal orientation on the insulating substrate, which influences such process parameters as etching speed, oxidation rate, and impurity diffusion coefficient, can only be controlled through the seeding process. In addition, one seeded epitaxial layer provides three active regions; the top and bottom regions of the epitaxial layer and the seeding region. These new regions should be effectively utilized to develop novel devices.

In line with this, the present paper describes the characteristics of elementary devices located at various regions in a seeded lateral epitaxial layer. Electrical properties are discussed in connection with the quality of the grown Si layer.

2. Seeded-Lateral Epitaxy on SiO₂

In the present experiments, a poly. Si layer (350 nm thick) was deposited on a Si substrate covered by SiO_2 (350 nm thick) stripe patterns.

Samples were then irradiated with a cw Ar laser (power: 2-8W, scan speed:5-100cm/sec, spot size:15-30μm) to melt and recrystallize the poly. Si films. When the laser beam was scanned at low power or high speed, no epitaxial growth was observed. With increased laser power or slowed scanning speed, poly. Si on the SiO_2 film tended to be stripped off. Thus, annealing in a narrow range was used to achieve uniform recrystallization in the wafer.

One example of epitaxially grown film (power:7w, scan speed: 50cm/sec) after Secco etching is shown in Fig. 1, where the laser beam scanned unidirectionally from left to right $(a \rightarrow b \rightarrow c \rightarrow d)$. Crystallinity after laser annealing was different in each region.

In region (a), poly. Si grain size was remarkably larger, however single crystal Si was not obtained. In region (b), the poly. Si turned into single Si through vertical epitaxial growth from the substrate, although high defect densities remained. This crystallization began to propagate in lateral directions at region(c) as the laser spot was moved. Consequently, single crystal Si about 15 μm long was obtained in region (d).

Crystal quality in the epitaxial film near the seeding area, i.e., the SiO_2 window area, was precisely determined using a TEM and a microprobe RHEED.[11] The results are shown in Fig. 2 (A) and (B). The diffraction patterns in Fig. 2 (A) confirm that the crystal orientation of the various parts of the grown layers are identical to the substrate Si.[12]

However, as indicated in Fig. 2(B), the TEM micrograph shows that dislocations were formed in the layer near the SiO_2 edge, corresponding to region (c) in Fig. 1. These dislocations escaped to the sample surface along the inclined plane of the SiO_2 edge during lateral growth. Consequently, no crystal defects were observed in region (d).

3. Influences of Seeding on Electrical Properties

Electrical properties of the grown layer were evaluated as a function

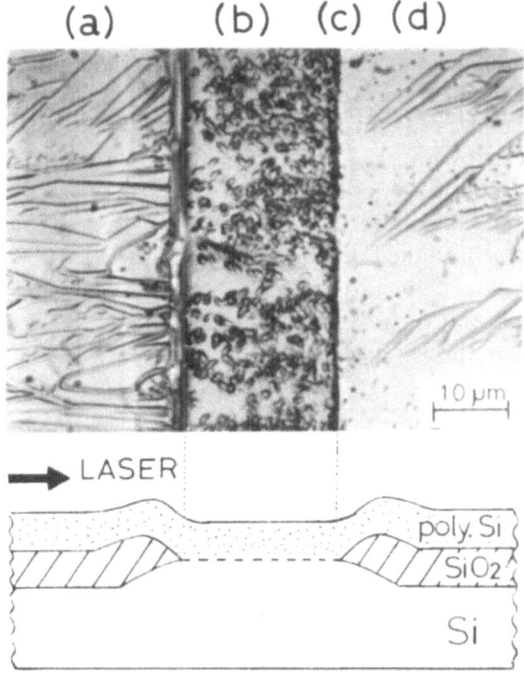

Fig. 1. Optical micrograph, and schematic cross section of laser annealed and Secco-etched poly. Si film on Si and SiO_2.

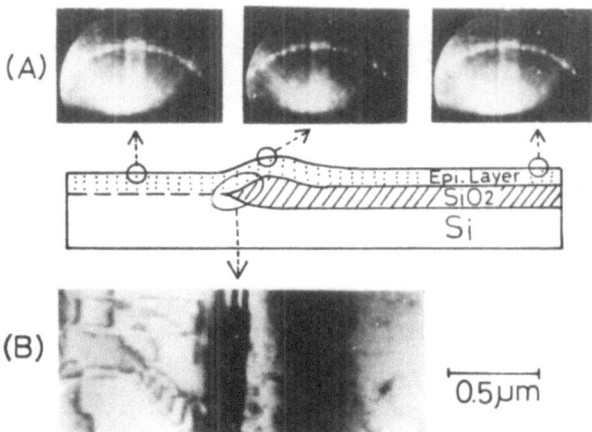

Fig. 2. (A) Electron diffraction patterns of the seeded-lateral epitaxial layer using a μ-RHEED. (B) TEM observation of the grown layer near the SiO_2 edge.

of distance from the seeding area, by fabricating n-channel MOSFET's.[13–16)]

Plan and cross-sectional views of the FET's (channel length: 8 μm, channel width: 15 μm) are shown in Fig. 3, where the gate regions for the FET's indicated as (a),(b),(c) and (d) are located in the corresponding epitaxial regions shown in Fig. 1. During fabrication, boron implantation $(6 \times 10^{11}$ cm^{-2}, 100 keV) and furnace annealing were used to reduce resistivities of the seeded lateral epitaxial layers. Source and drain regions were formed by self-aligned arsenic implantation $(1 \times 10^{16}$ cm^{-2}, 80 keV) to the poly. Si gate (350 nm) and furnace annealing (1000°C, 40 min). A surface photograph of a MOSFET fabricated on an SOI layer (case(d) in Fig. 3) is also shown in Fig. 4.

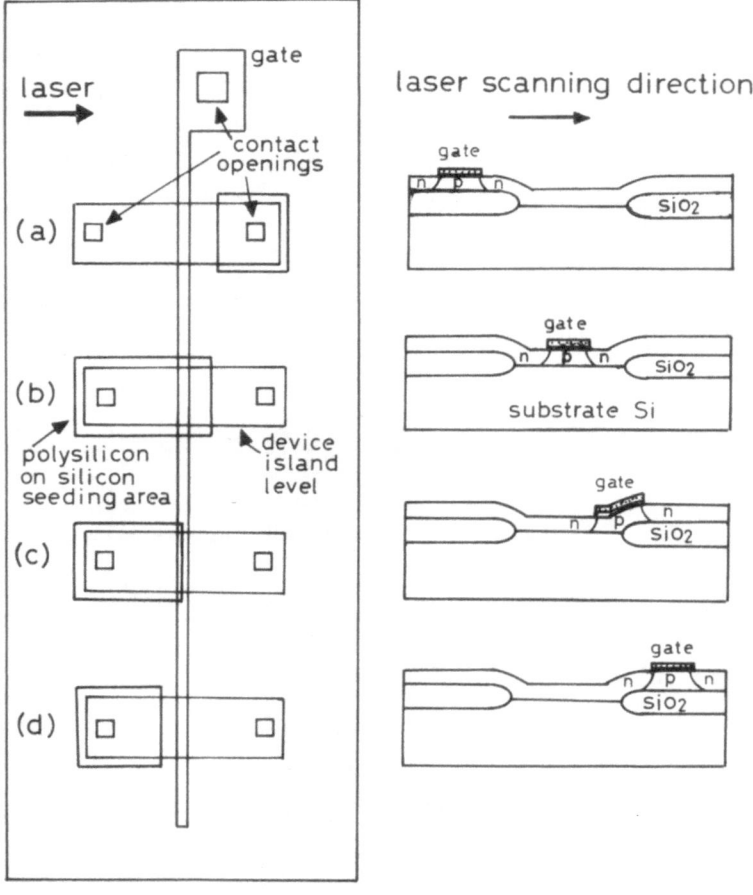

Fig. 3. Plan and cross sectional views of MOSFET's.

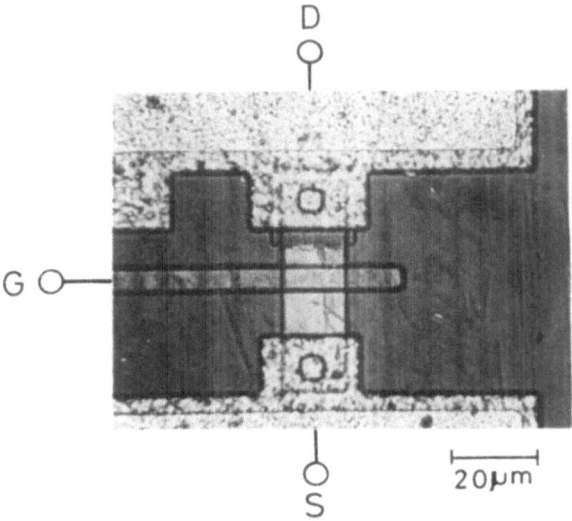

Fig. 4. Photomicrograph of a MOSFET fabricated on an epitaxial layer on SiO₂.

Drain current and voltage characteristics of typical samples are shown in Fig. 5. Channel mobilities derived from the *I-V* characteristics are summarized in Fig. 6 as a function of device distance from the seeding area.

In region (*a*), a dramatic increase in poly. Si grain sizes due to laser annealing improved electron mobility by an order of two. However, grain boundaries randomly distributed in this region (Fig. 1) cause scattering of electron mobilities over a wide range (100–500 cm²/V). Recent experiments[17] indicate that grain boundaries perpendicular to the current path decrease electron mobility. On the other hand, enhanced impurity diffusion along grain boundaries causes short circuiting between sources and drains; that is, grain boundaries parallel to the current path increase leakage current. These results clearly suggest that crystal quality in region (*a*) is still far from the LSI application level.

In region (*b*), mobility values of 300 cm²/V.s were obtained for samples without laser annealing, which suggests grain growth during furnace annealing. After laser annealing, high mobility (600 cm²/V.s) and reasonable threshold voltage (0.5 V) were obtained, although crystal defects still remained, as shown in Fig. 1. Recent observation of cross-sectional TEM[18] have indicated that dislocations are mainly localized near the interface between the deposited film and the substrate. These dislocations originate from contamination of the substrate surface before poly. Si deposition. This gives a good explanation for the large mobility value in the surface region.

M. Miyao *et al.*

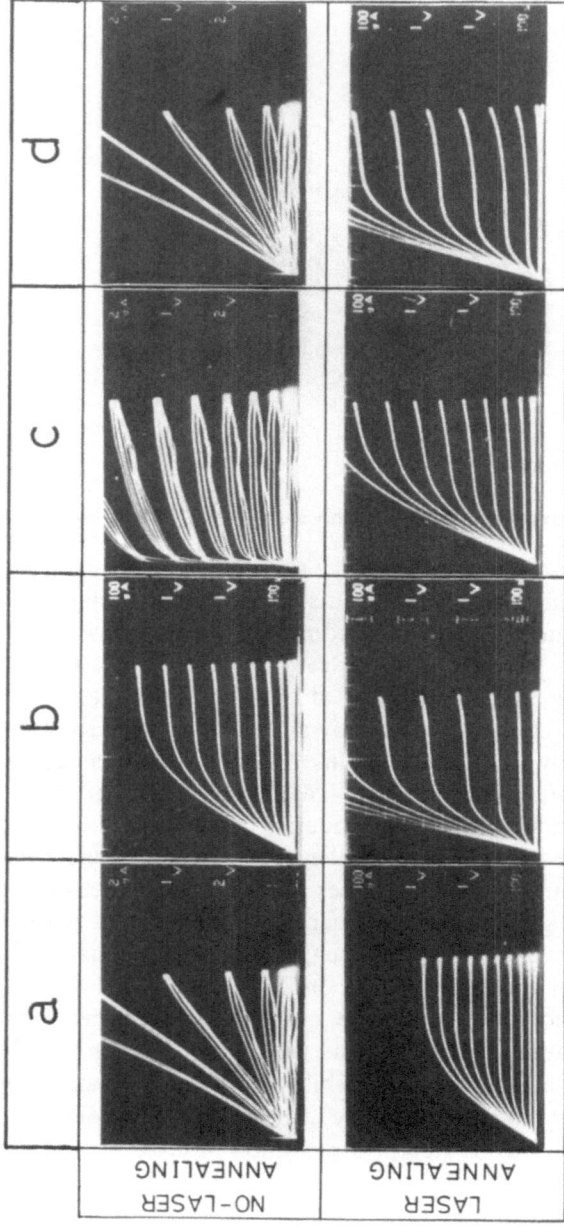

Fig. 5. *I-V* characteristics for typical MOSFET's. Results with and without laser annealing are compared. Vertical axis, horizontal axis and gate step scales are 2 μA, 1 V and 2V, respectively, for samples (a), (c) and (d) without laser annealing. They are 100 μA, 1 V, 1 V for (b) without laser annealing and for (a), (b), (c) and (d) with laser annealing.

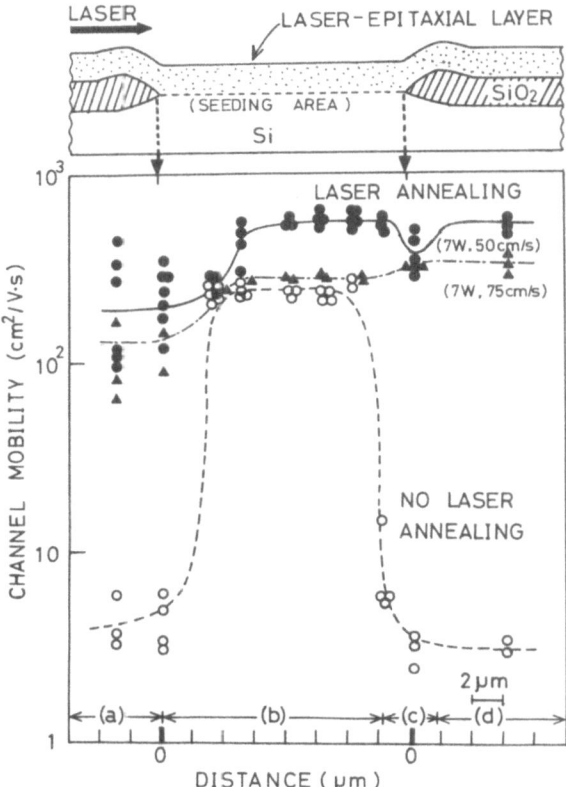

Fig. 6. Values of channel mobility summarized as a function of distance from the SiO₂ edge.

In region (c), the mobility value was 30% lower and threshold voltage was 0.5 V higher than for the other regions. TEM measurements indicated dislocations in this region also. In this case large differences in temperature rise and cooling rate during and after laser irradiation produce the residual stress discontinuities and could enhance dislocation formation. Such crystal degradation is the main reason for the low electron mobility and high threshold voltage.

In region (d), a representative mobility value identical to that for bulk crystal was obtained. Scattering of this mobility value was very small, unlike that in region (a). This result indicates that the seeding process is vital to realizing uniform crystallization in a wafer.

Electronic properties in the subthreshold region, shown in Fig. 7, indicate low leakage current (0.7–1.0 pA/micron-of-channel-width at $V_G = -5V$, $V_G = 5V$) and a high tailing factor (90mV/decade). The leakage

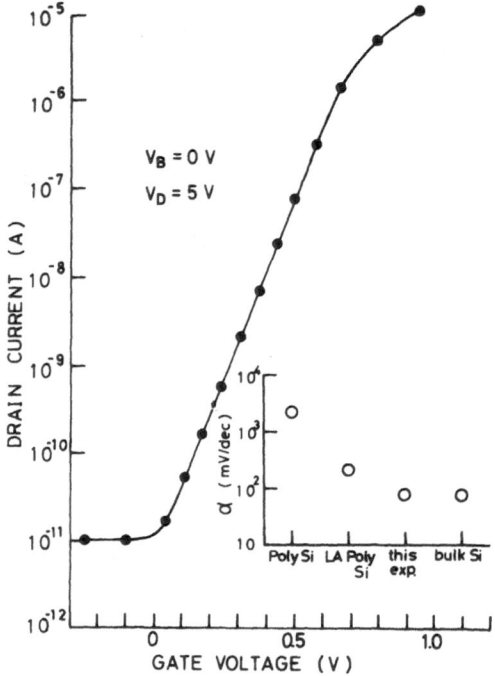

Fig. 7. Drain current and gate voltage characteristics in the subthreshold region for a MOSFET fabricated on an epitaxial layer on SiO_2. Tailing factor, α, is compared with the previously reported data.

current level is comparable to that for bulk Si, and is much smaller than that of published SOI data, [19] i.e., seeded lateral epitaxial Si on a Si_3N_4 substrate. This suggests that negative charge density at the bottom interface, which cause a serious problem when using a Si_3N_4 substrate, is very low when using a SiO_2 substrate. Experiments with back gate MOSFET's [15, 16] indicate that negative charge density at the grown layer and the oxide are in the low to mid 10^{11} cm^{-2} range.

4. Electrical Properties of Multiple-Regions in the Seeded Epitaxial Layer

To evaluate electrical properties of multiple-regions in the seeded lateral epitaxial layer, p-n diodes, front-gate MOSFET's and back-gate MOSFET's were fabricated. Electrical characteristics of these devices are shown in Fig. 8. The p-n junction characteristics in the seeding area (Fig. 8-B) indicate that the reverse current level is lower than 10^{-7} A/cm^2 at 0.1 V bias. The lowest value obtained is 6×10^{-8} A/cm^2. The ideality factor

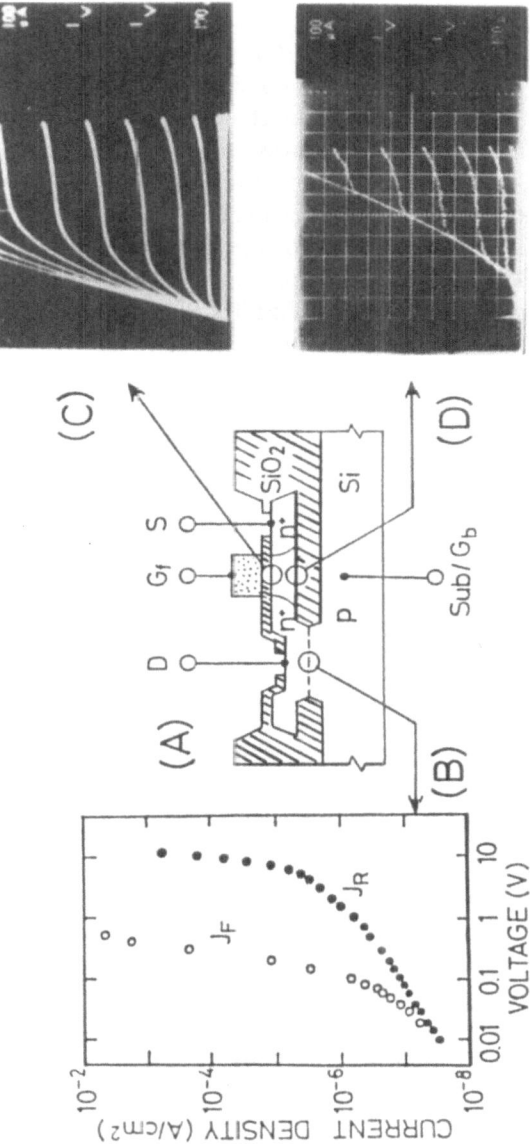

Fig. 8. (A) Schematic cross sections of seeded SOI devices. (B), (C) and (D) are *I-V* characteristics of a diode in the seeding area, a front gate MOSFET, and a back gate MOSFET, respectively.

in forward operation is in the range of 1.2–1.4.

These relatively poor characteristics are considered to be the results of residual defects at the interface between the Si substrate and the recrystallized layer. Thus, special cleaning of the Si substrate is necessary to improve the interface characteristics.

The conventional front-gate MOSFET (Fig. 8-C) indicates that electrical properties of the top region in the epitaxial layer are good, as was discussed in the previous section. The characteristics of the back-gate MOSFET, in which the poly. Si gate was shorted to the source and the Si substrate was used as the gate, is shown in Fig. 8-D. In this operation, a p-n diode is connected between the back-gate and the drain. Therefore, the forward biased *I-V* characteristics of the p-n diode are superimposed onto the original *I-V* characteristics, as shown in the figure.

Gain factor (g_m) of back-gate MOSFET's is summarized in Fig. 9-A as a function of laser scanning speed. The gain factor showed a maximum value ($50\mu\mho/V$) at a scanning speed of 30 cm/s. This value indicates ex-

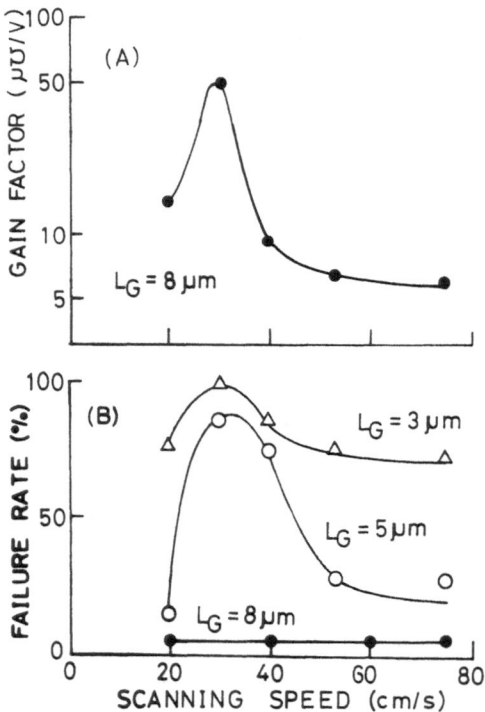

Fig. 9.   Gain factor (A), and failure rate (B) of back gate MOSFET's as a function of laser scanning speed.

tremely high apparent electron mobility (2000 cm^2/V.s). Apparent mobility, in this case, is defined as $g_m \cdot L_{mask}/C_{ox} \cdot W \cdot (V_G - V_{TH})$, where L_{mask}, W, C_{ox}, V_G and V_{TH} are gate length, gate width, gate capacitance, gate voltage and threshold voltage, respectively.

The failure rate for MOSFET's, i.e., the probability of source and drain being short circuited, is shown in Fig. 9-B. As shown by these figures, gain factor and failure rate showed maximum values at the same scanning speed. This suggests that side diffusion of arsenic atoms during furnace annealing becomes dominant at this annealing condition. Thus, abnormally high apparent electron mobility is concluded to be the result of the large difference between gate length (L_{mask}) and channel length ($L_{channel}$).

When scanning speed is decreased, the failure rate decreases and apparent electron mobility reaches a normal value. Moreover, it was found that a scanning speed of 20 cm/s corresponds to the threshold annealing condition needed to obtain single crystal growth over SiO$_2$.

These results suggest that insufficient melting during laser annealing causes stress or defects at the back interface, which enhances arsenic atom diffusion. Such an interface structure have not appeared in conventional Si process technology. A physical model of the interface between the molten Si and the insulator has yet to be developed.

5. Application to Gate Controlled p-n Diode

A novel device that utilizes this seeded lateral epitaxial structure has been fabricated. A schematic cross section and I-V characteristics for this device are shown in Fig. 10-(A) and (B), where channel region thickness, gate oxide thickness, buried oxide thickness, gate length and gate width

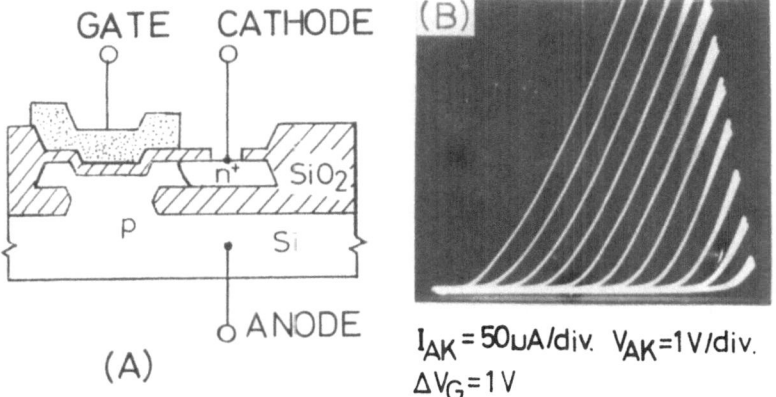

Fig. 10.   Schematic cross section (A), and I-V characteristics (B) of a gate-controlled p-n diode.

are 350nm, 53nm, 350nm, 8μm and 15μm, respectively. This device is essentially a lateral p-n diode [20] in which the junction barrier can be controlled by an external MOS gate. This is because the SOI layer thickness is thinner than the Debye length in the p-type region under the MOS gate. Therefore, under a certain gate bias, the forward p-n junction current can be cut off, even under high forward bias.

The anode to cathode *I-V* characteristics were measured as a function of gate bias. They show triode-like behavior, with the forward *I-V* characteristics of the diode shifting along the cathode bias axis as gate bias changes. This device has the inherent high input impedance of a MOS gate, and a low output impedance due to the bipolar action of the device.

6. Conclusions

The detailed electrical characteristics of a seeded lateral epitaxial Si layer have been described. The results obtained from MOSFET and diode fabrication indicate that the crystal quality of such Si layers on SiO_2 is sufficent for LSI application. This strongly indicates the usefulness of the seeding process in SOI fabrication. Possible application of multiple-regions in an epitaxial layer to novel device structure was proposed.

However, some problems, mainly concerned with crystal growth at the seeding area and SiO_2 edge, remain. Si deposition in an UHV after special cleaning of the Si substrate should improve the interface characteristics. In addition, calculation of temperature profiles and cooling rate are effective in determining optimized sample structures. Investigation of the physical structure and properties of the back interface is also important from both fundamental interest and device application viewpoints.

Acknowledgements
The authors would like to thank Drs. I.Takemoto, N.Hashimoto and H.Sunami for their help in device fabrication. Drs. M.Tamura, M.Ichikawa, N.Natsuaki and T.Warabiasako are also to be acknowledged for the TEM and micro-probe RHEED observations and for their useful discussions.

REFERENCES

1) *Laser and Electron Beam Processing of Materials*, ed. C.W.White and P.S.Peercy (Academic Press, New York, 1980).
2) *Laser and Electron Beam-Solid Interactions and Material Processing*, ed. J.F.Gibbons, L.D.Hess, and T.W.Sigmon (North-Holland, New York, 1981).
3) *Laser and Electron-Beam Interactions with Solids*, ed. B.R.Appleton and G.K.Celler (North-Holland, New York, 1982).
4) J.F.Gibbons, K.F.Lee, T.J.Magee, J.Peng, and R.Ormond: Appl.Phys.Lett. **34** (1979) 831.

5) A.F.Tash, T.C.Holloway, K.F.Lee, and J.F.Gibbons: Elect.Lett. **15** (1979) 435.
6) M.Tamura, H.Tamura, and T.Tokuyama: Jpn.J.Appl.Phys. **19** (1980) L23.
7) M.Tamura, H.Tamura, M.Miyao, and T.Tokuyama: Jpn.J.Appl.Phys. **20** (1981) Suppl.20-1.
8) M.Tamura, M.Ohkura, and T.Tokuyama: Jpn.J.Appl.Phys. **21** (1982) Suppl.21-1.
9) H.W.Lam, R.F.Pinizzotto, and A.F.Tasch Jr: J.Electrochem. Soc. **128** (1981) 1981.
10) H.I.Smith and D.C.Flanders: Appl.Phys.Lett. **32** (1978) 349.
11) M.Ichikawa and K.Hayakawa: Jpn.J.Appl.Phys. **21** (1982) 145.
12) M.Ohkura, M.Ichikawa, M.Miyao, and T.Tokuyama: Appl.Phys.Lett. **41** (1982) 1089.
13) M.Miyao, M.Ohkura, I.Takemoto, M.Tamura, and T.Tokuyama: Appl.Phys.Lett. **41** (1982) 59.
14) M.Miyao, M.Ohkura, I.Takemoto, M.Ichikawa, M.Tamura, and T.Tokuyama: *Proc. 14th Conf. (1982 Int.) Solid State Device, Tokyo, 1982:* Jpn J.Appl.Phys. **22** (1983) Suppl.22-1.
15) M.Miyao, M.Ohkura, T.Warabisako, and T.Tokuyama: Laser-Solid Interactions and Transient Thermal Processing of Materials (Nov. 1982, Boston).
16) T.Warabisako, M.Miyao, M.Ohkura, and T.Tokuyama: International Electron Device Meeting, (Dec. 1982) p433.
17) K.K.Ng, G.K.Cell, E.I.Povilonis, R.C.Frye, H.J.Leamy, and S.M.Sze: IEEE Electron Device Lett. **2** (1981) 316.
18) T.I.Kamins, T.R.Cass, C.J.Dell'Oca, K.F.Lee, R.F.W.Pease, and J.F.Gibbons: J.Electrochem.Soc. **128** (1981) 1151.
19) H.W.Lam: *Laser and Electron-Beam Interactions with Solids*, ed. B.R.Appleton and G.K.Celler North-Holland, New York, 1982).
20) Y.Omura: Appl.Phys. Lett. **40** (1982) 528.

Silicon-on-Insulator: Its Technology and Applications, edited by S. Furukawa, pp. 283–294.
© KTK Scientific Publishers, Tokyo, 1985.

DEVICE APPLICATION OF SIMOX (Separation by IMplanted OXygen) STRUCTURE

K. Kajiyama, K. Izumi, and S. Nakashima

*Electrical Communication Laboratories, Nippon Telegraph and Telephone
Public Corporation, Atsugi, Kanagawa 243-01, Japan*

Abstract SIMOX/SOI is produced by high-dose ($\sim 1 \times 10^{18}$/cm^2) and high-energy (~ 100keV) oxygen implantation, forming a buried oxide layer and reserving the surface single-crystalline Si layer. Annealing improves surface Si crystallinity and buried Si–O bonding. Epitaxial growth adds necessary Si thickness and improves crystallinity further. Using SIMOX, a 1kb CMOS static RAM and a 400-gate PLL IC have been successfully fabricated. They proved to have good electrical characteristics and stable operation. SIMOX is good for VLSIs, because the fabrication process is reproducible and circuit design is flexible. A higher-current implanter is desired.

1. Introduction

The single-crystalline Si on insulator (S-SOI) structure is promising for CMOS VLSIs used in high-speed and low-dissipation-power operation with high packing density. Reported S-SOIs are classified into three groups: growing single-crystalline Si on insulators (type S), burying the insulator layer in single-crystalline Si (type I), and turning the insulator-on-silicon structure upside down (type V). Among the many S-SOI structures available, only several have been successfully used in fabricating LSIs, for example, SOS (Silicon On Sapphire),[1] SIMOX,[2] FIPOS (Full Isolation by Porous Oxidized Silicon)[3] and CEPIC (Complementary Epitaxial Passivated Integrated Circuit).[4]

SOS is in the "S" category. In that, Si is grown hetero-epitaxially on a sapphire substrate. SOS-LSIs are commercially available from pertinent suppliers. SIMOX and FIPOS are in the "I" category. To form the buried oxide, SIMOX utilizes uniform oxygen implantation, while FIPOS utilizes local anodization and thermal oxidation. A 1kb CMOS static RAM (Random Access Memory) and a 400-gate PLL (Phase Locked Loop) IC have been fabricated with SIMOX, while a 16kb CMOS static RAM has been fabricated with FIPOS. CEPIC is in the "V" category, in which a thick-polycrystalline-Si/oxide-layer/substrate-Si structure is turned upside down and the substrate is then lapped to leave single-crystalline Si islands. CEPIC has been successfully used in fabricating bipolar LSIs for a subscriber line interface circuit containing 500 n-p-n and p-n-p transistors, including 100 high-voltage (350V) transistors.

SIMOX has various advantages in regard to LSI fabrication. Implantation results in good reproducibility and promises high yields for large-component-number devices. The uniform oxide layer results in a flat surface, and assures fine lithography for small devices and large flexibility for circuit design. Implanted oxygen does not contaminate Si, unlike Al in SOS. In addition, dry processing saves resources. However, SIMOX has a disadvantage in that implantation takes a longer time than other LSI processes. The oxygen beam-current trend is discussed in Section 5.

Current status of SIMOX is reviewed in the following.

2. Simox Fabrication Process

The SIMOX is fabricated simply by three major steps, implantation, annealing and epitaxial growth (Fig. 1). Wafer warpage is the same as that of thermally oxidized wafer and the epitaxial layer is shiny. In the present experiment, Si wafer orientation is usually (100) and, for reference, (111).

2.1 Implantation

Oxygen implantation forms a buried oxide layer, preserving the single-crystalline Si layer on the buried oxide, on conditions that dose is high ($\sim 1 \times 10^{18}$/cm^2), implantation energy is high (~ 100keV) and wafer

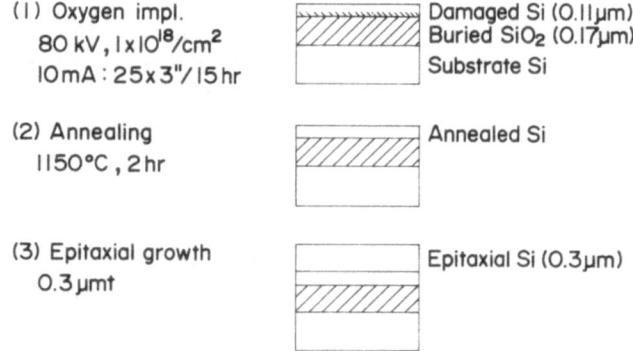

Fig. 1. SIMOX fabrication process.

temperature is high ($\sim 400°C$). Hot implantation recovers implantation-induced defects during implantation, otherwise surface Si becomes amorphous owing to the high dose. In the present experiment, wafers were not heated intentionally but were heated by the ion beam itself.

Depth profiles for implanted oxygen change from Gaussian to trapezoidal, as dosage increases (Fig. 2).[5] Profiles were measured by RBS (Rutherford Backscattering Spectroscopy), showing that the peak oxygen/silicon (O/Si) ratio saturates near the stoichiometric ratio ($= 2.0$, for SiO_2). The saturated O/Si ratio approaches 2.0, as implantation current increases, i.e. as wafer temperature rises. Saturation starts near $1.0 \times 10^{18}/cm^2$ does at 80keV energy and near $1.7 \times 10^{18}/cm^2$ dose at 150keV energy. A lower energy decreases the saturation dose and the surface Si thickness.

However, excessively low energy fails to preserve single-crystalline Si surface. Minimum implantation energy for preservation depends on implantation conditions, especially on wafer temperature during implantation. A 70keV energy was experimentally proved to be sufficient to reserve crystallinity for epitaxial growth, where dose is $0.90 \times 10^{18}/cm^2$ (near the saturation dose) and current is $300\mu A$ (converted from $O_2{}^+$- to O^+- implantation*) in 2cm×2cm. Usually, in the present experiment, oxygen was implanted with $1.0 \times 10^{18}/cm^2$ at 80keV or $1.2 \times 10^{18}/cm^2$ at 150keV.

Implanted oxygen atoms migrate from oxygen-sufficient to oxygen-deficient regions during implantation. The internal oxidation produces an abrupt interface between the surface Si and the buried oxide. The transient layer is thinner than 20nm as measured by AES (Auger Electron Spectroscopy). The interface electrical characteristics are similar to those for thermal oxidation, as shown in Section 3.

On the contrary, however, in nitrogen implantation, the nitrogen to Si ratio does not saturate, and excess nitrogen atoms stay in free or weakly-bonded states (Fig. 3)[5] The RBS profile remains Gaussian-like and the peak value does not saturate. In the XPS (X-ray Photo-electron Spectroscopy) profile, the peak value saturates near the stoichiometric ratio ($= 4/3$, for Si_3N_4). In XPS measurement, free or weakly bonded nitrogen atoms escape during Ar sputtering to reveal depth profile.

2.2 Annealing

Annealing improves surface-Si crystallinity and strengthens Si–O bonding after the implantation. Still single-crystalline Si is epitaxially grown

---

*In the present experiment, $O_2{}^+$ was implanted under the condition that the dose is 0.45×10^{18} O_2/cm^2, energy is $140keV/O_2$ and current is $150\mu A$ in 2cm×2cm. This condition is equivalent to O^+-implantation condition that the dose is 0.90×10^{18} O/cm^2, energy is $70keV/O$ and current is $300\mu A$ in 2cm×2cm.

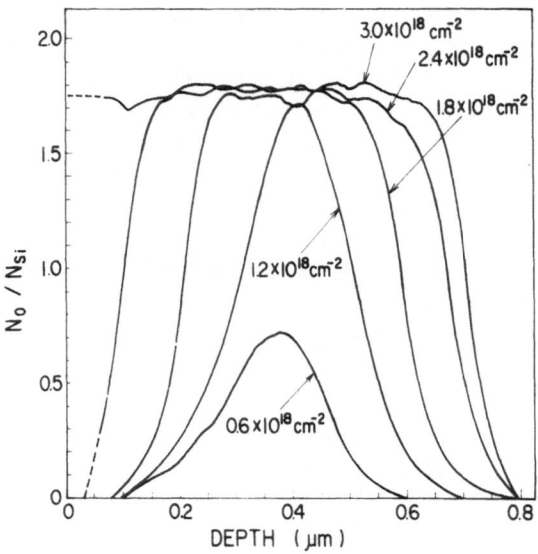

Fig. 2.   Depth profiles for O/Si atomic ratio. Wafers are (111) 3″φ Si. Implantation energy
was 150keV and current was 100μA in 2cm × 2cm. Profiles were measured by RBS.

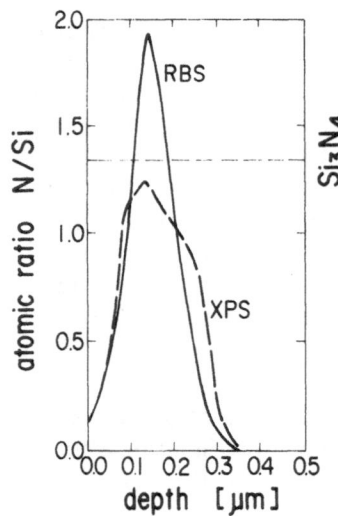

Fig. 3.   Depth profiles of N/Si atomic ratio. Wafer was (111) 3″φ Si. Implantation energy
was 70keV, dose was 1.2×10^{18}/cm² and current was 300μA in 2cm × 2cm (converted from N_2^+-
to N^+-implantation. Profiles were measured by RBS and XPS.

on the as-implanted wafer and Si–O bonding is observed in the as-implanted oxide, without annealing. In annealing, oxygen depth profiles change only in the small concentration region.

A higher temperature and a longer exposure time are desirable for improvement. Crystallinity improvement is effective before and even after the epitaxial growth. Usually in the present SIMOX process, wafers are annealed at 1150°C for 2 hours in N_2 ambient before epitaxial growth.

Crystalline defects in the (100) epitaxial layer (epitaxial condition is predicted in Section 2.3) are markedly reduced by annealing at 1400°C (Fig. 4). Defects were investigated with etch-pit density (EPD) after etching 0.2μm-deep for the 0.3μm-thick CVD epitaxial layer, utilizing the Wright solution (HF: 60cc, HNO_3: 30cc, CrO_3: 5mol in 60cc H_2O, $Cu(NO_3)_2$: 2g, CH_3COOH: 60cc, H_2O: 60cc). Etch-pits are mainly due to dislocations. TEM (Transmission Electron Microscope) observation reveals dislocation networks.[6]

As-epitaxial wafers are processed by the conventional resistance-heater furnace annealing (FA) and infrared radiation annealing (IRA).[7] In the present experiment, the authors did not amorphize the interface between the surface Si and the buried oxide prior to annealing, unlike in the SOS[8] case. FA raises wafer temperature only to 1200°C (near the fused-silica softening temperature). On the other hand, IRA raises wafer temperature to 1400°C (just below the Si melting point). EPD decreases as annealing

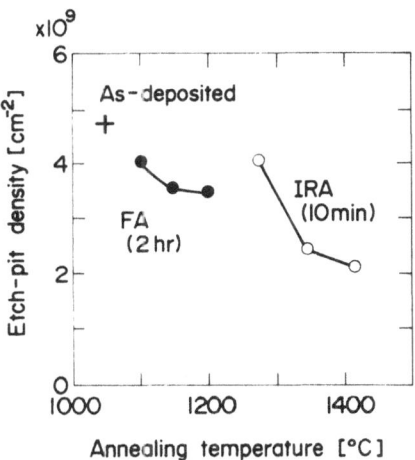

Fig. 4.   Temperature dependence of etch-pit density. Oxygen atoms were implanted with 1×10^{18}/cm^2 dose at 80keV energy. As-implanted wafers were previously annealed for 2 hours at 1150°C in N_2 ambient with a conventional resistance-heater furnace. Epitaxial thickness is 0.3μm. Below 1200°C, FA is utilized for 2 hours. Above 1250°C, IRA is utilized for 10 minutes. Annealed wafers are etched to 0.2μm depth with Wright solution.

temperature rises. EPD increase, as the present annealing temperature changes from 1200°C to 1250°C, because annealing time is shortened from FA (2 hours) to IRA (10 minutes). IRA overheats for long-time annealing, while FA fails short-time annealing. IRA at 1400°C reduces EPD to $2 \times 10^9/cm^2$ (40% of the as-epitaxial). This EPD is lower than that in a commercial SOS with the same $0.3\mu m$ epitaxial thickness, while the SOS etch-pits are mainly due to stacking faults.

Annealing decreases dangling bonds. As annealing temperature is higher, ESR signal intensity becomes lower both in the surface Si ($g = 2.006$, coinciding with that in amorphous Si), and in the buried oxide ($g = 2.002$, coinciding with that in CVD oxide).[9]

Si–O bonding is weak in the as-implanted oxide and approaches that for thermal oxide through annealing. Si–O bonding is measured by infrared absorption, refractive index, dielectric constant and breakdown electric field in the trapezoidal-profile samples without the surface Si layer (Table 1). The authors consider the Si–O bonding in the implanted surface-oxide as being the same as that in the buried oxide.

2.3 Epitaxial Growth

Epitaxial growth adds the necessary Si thickness and improves crystallinity as well. Naturally, devices are possible to fabricate using SIMOX without the epitaxial growth, if implantation energy is high (for example, 150keV). In the present experiment, epitaxial Si was deposited by a conventional CVD process with SiH_4/H_2 at 1050°C.

Table 1. Si–O bonding of the implantation-formed oxide in the as-implanted and annealed samples, compared with the conventional thermal oxide.

Sample	As-implanted 25 keV, $7 \times 10^{17}/cm^2$ $80 \mu A/(2 cm \times 2 cm)$	Annealed 1150°C, 2 hr in N_2	Thermal oxide 1100°C in dry O_2
Infrared absorption			
Peak	$1040 cm^{-1}$	$1070 cm^{-1}$	$1070 cm^{-1}$
FWHM	$10 cm^{-1}$	$7 cm^{-1}$	$10 cm^{-1}$
Refractive index ($\lambda = 0.63 \mu m$)	1.51	1.47	1.46–1.47
Dielectric constant ($f = 1 MHz$)	4.2–4.3	3.9–4.2	3.8
Breakdown field	$4–7 \times 10^6 V/cm$	$6–9 \times 10^6 V/cm$	$7–9 \times 10^6 V/cm$

Heat treatment in H_2 ambient does not degrade buried oxide and changes oxygen profiles only in the low concentration region. Dangling bonds were decreased favorable in H_2 ambient, even at a lower temperature compared to that in N_2 ambient.[9]

3. Electrical Characteristics

SIMOX shows good electrical characteristics for the surface Si, the buried oxide and the interface between them.

The electron and hole mobility in the surface Si is as high as for bulk Si (Fig. 5). Mobility was measured by MOS FET transconductance in 0.3μm-thick surface-Si layer with a Gaussian-like oxygen profile (80keV, $1.0 \times 10^{18}/cm^2$). Doping is $6 \times 10^{14}/cm^3$ for n-islands and $3 \times 10^{16}/cm^3$ for p-islands.

Breakdown field intensity of the buried oxide is the same as that for thermal oxide (Table 1). In order to fabricate high-voltage devices, a thicker buried-oxide layer is necessary. Accordingly, higher dose and higher energy were utilized. For example, the authors have fabricated n-MOS FETs for 180V drain breakdown voltage and p-MOS FETs for 250V drain breakdown voltage, utilizing $1.8 \times 10^{18}/cm^2$ dose and 150keV energy.[10]

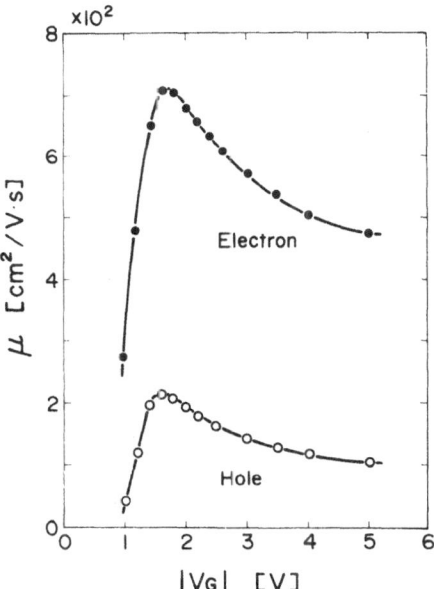

Fig. 5. Field effect mobility in 0.3μm-thick Si for SIMOX with 80keV energy and $1.0 \times 10^{18}/cm^2$ dose. Mobility is measured for electrons with n-MOS FET and for holes with p-MOS FET.

When oxygen profile is trapezoidal, the interface is similar to that for the thermal oxide. The flat-band interface charge density $Q_{fb}/q = 8 \times 10^{10}/cm^2$ at 1MHz (frequency dependence is small in the 2kHz–10MHz range) and the minimum interface trap density in the bandgap $D_{it} = 1 \times 10^{11}/cm^2$ (Fig. 6).[11] Interface characteristics were measured by an inverted MOS diode utilizing the buried oxide as a gate-oxide layer. As Q_{fb}/q and D_{it} are small in the trapezoidal profile, the surface-Si Fermi level is affected by substrate-Si potential and LSI circuit design becomes complicated in operating condition.

On the contrary, when the oxygen profile is Guassian-like, the interface serves as a passivation layer and lateral leakage current is 1pA for a 10µm-width lateral p-n junction at 10V reverse-bias. The leakage is smaller than in conventional SOS. The interface transient layer is a mixture of polycrystalline Si and oxide, like SIPOS[12] (Semi-Insulating Polycrystalline-Silicon). Its resistivity is high enough to reduce leakage current and its trap density is high enough to shield the electric field.

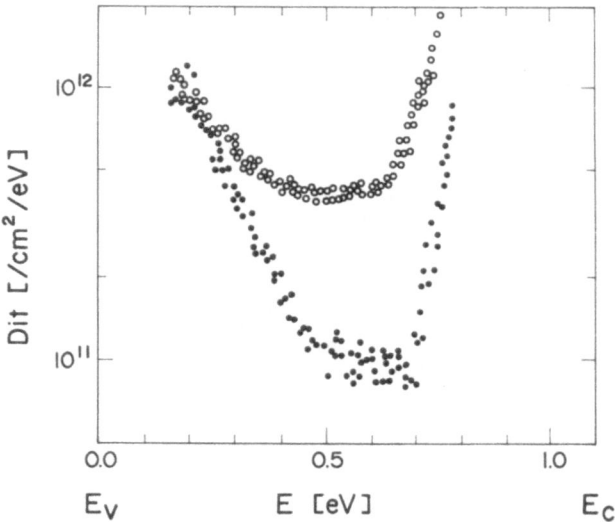

Fig. 6. D_{it} distribution in the bandgap. •: $2.4 \times 10^{18}/cm^2$ (trapezoidal profile) and ○: $1.8 \times 10^{18}/cm^2$ (non-trapezoidal profile) at 150keV. D_{it} is calculated from the quasi-static C-V curves measured at gate bias from $-20V$ to 15V with 100mV/s ramp rate. Sample preparation condition is as follows. Oxygen is implanted into n($\sim 10^{16}/cm^3$, 1.4µm, undoped epitaxial layer)/n$^+$($\sim 10^{19}/cm^3$, 1µm, As-diffused layer)/n$^-$($1 \times 10^{15}/cm^3$, (100)-orientation substrate)-Si. Wafer was annealed at 1150°C for 2 hours in N$_2$ ambient and an additional epitaxial layer (1.4 $\times 10^{16}/cm^3$, 1.5µm) was grown on it. After upper contact formation (P-implantation/drive and Al-deposition), wafer was annealed at 400°C for 30 minutes in H$_2$ ambient.

Consequently, the authors prefer the Gaussian-like profile to the trapezoidal profile in LSI fabrication.

4. *Application to LSIS*

The authors have succeeded in fabricating a 1kb CMOS static RAM and a PLL frequency synthesizer IC using a Gaussian-like oxygen profile SIMOX with $1.0 \times 10^{18}/cm^2$ dose and at 80keV energy (Fig. 7).[2] In 0.3μm-thick epitaxial layer, component MOS FETs were electrically isolated by buried oxide in the vertical direction and by conventional LOCOS (Local Oxidation of Silicon) oxide in the lateral direction. Because LOCOS oxide is readily connected with SIMOX oxide, Si islands are fully isolated from each other with oxide, unlike SiO_2/Al_2O_3 in SOS. The circuit design-rule was 2μm, effective channel length was 1.5μm and gate-oxide thickness was 50nm.

The RAM chip size is 2.3mm $\times$ 2.3mm. It contains about 7000 MOS transistors. The memory cell is a 6-transistor flip-flop in 40μm $\times$ 35μm area. The resultant chip-select access-time is 12ns and dissipation power is 45mW at a 5V supply voltage. Access time is 30% smaller than for conventional non-SOI circuits.

PLL chip size is 1.0mm $\times$ 1.7mm. It contains 400 gates. Maximum operation frequency is 420MHz. Dissipation power is 65mW at a 5V supply voltage.

Fig. 7.   Chip micro-graph for 1kb CMOS static RAM using SIMOX. Chip size is 2.3mm $\times$ 2.3mm.

The operation condition does not change for several hours, confirming that the surface-Si/buried-oxide interface is stable.

5. Discussion and Conclusion

To increase SIMOX through-put, the authors are raising implanter oxygen beam current (Fig. 8). The authors are now using a 10mA and 80keV implanter equipped with a microwave ion-source. Implantation through-put is 25 3″φ wafers per 15 hours. The microwave ion-source is used for more than 400mA·hour continuous operation, though the conventional hot-filament ion-source is degrated by oxygen within 100mA·hour operation. The authors are planning an implanter with a higher beam current and a higher accelerating voltage.

In the course of a preliminary study for three-dimensional LSIs, the authors fabricated a triple S-SOI structure utilizing a three-cycle SIMOX process (Fig. 9).[13] The top surface is single-crystalline, according to the RBS channeling spectrum and electron diffraction pattern. As the present SIMOX process needs high temperature processing, the triple S-SOI does not contain any devices. However, the success of the triple S-SOI gives assurance that SIMOX can be easily reproduced and, promises to have good application to VLSIs.

The present SIMOX is suitable for CMOS FETs, such as static RAMs

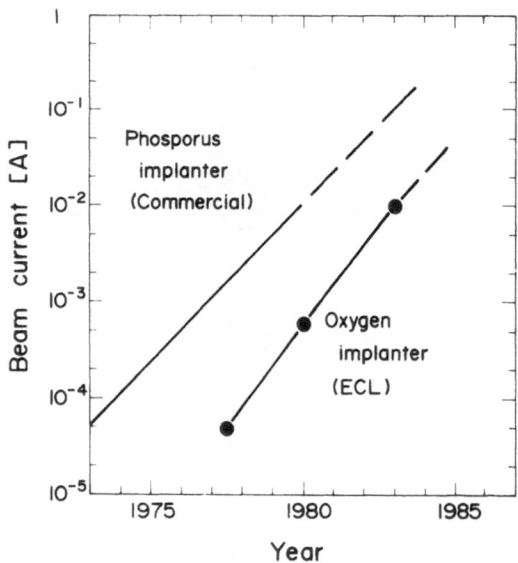

Fig. 8.   Beam current trend in ECL (Electrical Communication Laboratories) oxygen implanter and commercial phosphorous implanter.

Fig. 9.   Triple S-SOI formed by three-cycle SIMOX process. Cross sectional view of a cleaved and etched (dilute HF) surface by SEM (Scanning Electron Microscope). Shiny area is single-crystalline Si and dark stripes are buried oxide layers. The white broken line is $1\mu m$ long. Implantation: 80keV, $1.0 \times 10^{18}/cm^2$. Annealing: 1150°C, 2 hours, in N_2 ambient. Epitaxial layer: $1\mu m$ thick.

and logics, due to LSI processing simplicity and circuit design flexibility, but not for dynamic RAMs or bipolar devices, due to high-density dislocations. A higher current implanter will promote further SIMOX application.

Acknowledgements

The authors would like to thank Mr. Y. Irita for oxygen implantation and characterization, and Dr. T. Suzuki, Mr. H. Ariyoshi, Mr. H. Ikawa and Mr. T. Sakai for encouragement.

REFERENCES

1)   Y. Okuto, M. Fukuyama, and Y. Ohno: IEEE J. SC **17** (1982) 204.
2)   K. Izumi, Y. Omura, M. Ishikawa, and E. Sano: 1982 Symp. VLSI Tech. Oiso, Digest Tech. Papers, p. 10.
3)   T. Mano, T. Baba, H. Sawada, and K. Imai: 1982 Symp. VLSI Tech. Oiso, Digest Tech. Papers, p. 12.
4)   T. Sakurai, T. Ohno, K. Kato, Y. Inabe, and T. Hayashi: IEEE Trans. ED **30** (1983) 1278.
5)   S. Maeyama and K. Kajiyama: Jpn. J. Appl. Phys. **21** (1982) 744.
6)   Y. Homma, M. Oshima, and T. Hayashi: Jpn. J. Appl. Phys. **21** (1982) 890.
7)   R. Komatsu and K. Kajiyama: J. Appl. Phys. **55** (1984) in print.
8)   J. Amano and K. W. Carey: J. Cryst. Growth **56** (1982) 296.
9)   M. Tabe and Y. Omura: 1982 Spring Meeting of Appl. Phys. Soc. (Japan), Abstract p. 621.
10)   M. Akiya, S. Nakashima, and K. Kato: *Proc. 14th Conf. Solid State Devices, Tokyo,*

1982, Jpn. J. Appl. Phys. **22** (1983) Supppl. 22-1, p 85.

11)   S. Nakashima and K. Ohwada: Jpn. J. Appl. Phys. **22** (1983) ll19.

12)   H. Mochizuki, T. Aoki, H. Okayama, M. Abe, and T. Ando: *Proc. 7th Conf. Solid State Devices, Tokyo, 1975*, Jpn. J. Appl. Phys. **15** (1976) Suppl. 15-1, p 41.

13)   Y. Irita, Y. Kunii, M. Takahashi, and K. Kajiyama: Jpn. J. Appl. Phys. **20** (1981) L909.

AUTHOR INDEX